Patrons, Curators, Inventors and Thieves

Patrons, Curators, Inventors and Thieves

The Storytelling Contest of the Cultural Industries in the Digital Age

Jonathan Wheeldon

First published 2014 by
PALGRAVE MACMILLAN

Palgrave Macmillan in the UK is an imprint of Macmillan Publishers Limited,
registered in England, company number 785998, of Houndmills, Basingstoke,
Hampshire RG21 6XS.

Palgrave Macmillan in the US is a division of St Martin's Press LLC,
175 Fifth Avenue, New York, NY 10010.

Palgrave Macmillan is the global academic imprint of the above companies
and has companies and representatives throughout the world.

Palgrave® and Macmillan® are registered trademarks in the United States,
the United Kingdom, Europe and other countries.

ISBN 978–0–230–24943–1

A catalogue record for this book is available from the British Library.

A catalog record for this book is available from the Library of Congress.

Transferred to Digital Printing in 2014

To Fiona,
for making everything worthwhile

Contents

Figures

Introduction: A Changing Master-Narrative of Cultural Production

Let me make the songs of a nation, and I care not who makes its laws.

(Plato)[1]

Written today, this maxim might also include the novels, soap operas and movies of a nation, but the point would still be well made. The cultural industries attract more sociological and political attention than their financially measured size would otherwise justify. This is because their primary concern is the production of social meaning. They can simultaneously reinforce and disrupt our perceptions of reality: of what is good and bad; right and wrong; relevant and irrelevant; fair and unfair. Compared with the clearer functional aims of most other industries, they have a greater influence on the interpretations of our complex and contested world, and on how we should engage with it. In this way, their impact on behaviour can be deeper and more subtle than that achieved through laws and regulations.

Culture only exists as an 'industry' of notable size and concentrated power because of continual technological innovation in the creation, reproduction and dissemination of texts, sounds and images. From Gutenberg to Sony, the commercializers of technology and of culture have historically collaborated in a more or less virtuous circle and, until 1999, innovation had never made the cultural industries poorer. I was tempted by alliterative impulse to write 'Gutenberg to Google', but stopped short by a few years because something changed in the nineties. It was something which made subsequent commercializers of technology less empathic towards the cultural industries. The development of the Web, the compression technologies collectively known as MP3, and peer-to-peer file-sharing were driven mostly by open collaboration in the pursuit of efficiency in communications, and in knowledge

1

discovery and cultural dissemination for their own sake, rather than by a commercial vision of how much money could be made from them. This gentle snub to the traditional mechanisms of capitalism quietly disrupted all the institutional rules of engagement in the early years of their evolution, and had unforeseen consequences once their significance was realized.

The institutional rules thus disrupted are underpinned by intellectual property law. Intellectual property law is the previously low-profile back-office mechanism which divided up the cultural industries' pie, reinforcing the worth and social identities of the key players, and keeping authors, artists, technologists, innovators, entrepreneurs and financiers more or less in commercial harmony for three centuries. These laws are no longer the exclusive conversational domain of lawyers, talent agents and accountants. They have emerged as a topic of heated public debate, which has now been running for well over ten years, without significant legal reform. Copyright law in particular has been thrown into the political spotlight and is suspected to be no longer fit for purpose. The problem is that products of innovation, once they become industrialized, institutionalized and embedded in the social identities and common sense of their age, have a tendency to breed fear and mistrust of further innovations; especially those which are beyond the control of the dominant players, and especially where the social rules of engagement have changed. This is the political dilemma of the digital age.

In the broadest sense, this book is concerned with how the social world of human beings is organized more by narrative plausibility than precision; how mastery of language is often more valuable than the control of economic assets, and how provisional and contextual sense-making, rather than scientific principle, tends to drive economic policy and strategic decision-making. With a fleeting nod to the storytelling contest of Chaucer's *Canterbury Tales,* it recognizes that identity, authority, power and structure in society can be both reinforced *and* critiqued through familiar tales of protagonists and antagonists, heroes and villains. The patron, the inventor and the curator (in various 20th century guises) are the traditional protagonists of cultural production. Their presumed expertise in filtering and nurturing drives an economic system of cultural intermediation which is as lucrative as it is influential. There are a number of equally familiar antagonists which undermine the authority and power of the cultural producers. Foremost amongst these antagonists is the intellectual property 'thief', who has been alternately vilified and romanticized for centuries with the colourful and ambivalent label of the 'pirate'. Though these competing narratives are not new, the digital age gives birth to some new, more nuanced characters.

The empirical roots of the book are in the music industry and its fascinating new millennium transformation through the seismic shifts in technology, communications and related social behaviour. Of most interest is the nature of the strategic struggle faced by corporate managers, and by government intellectual property policy-makers, to deal with this transformation. There is already plenty of interesting literature on the 21st-century challenges to the music industry in blogs, books and journals, and this book is by no means an introduction, nor a comprehensive overview. It is targeted at those who seek fresh and alternative ways to approach the complexity of the topic. My particular contribution has two original dimensions. Firstly, I take a critical, language-based approach and concentrate on the extent to which change-behaviour is empowered, but mostly constrained, by deep-rooted, identity-bound, social narratives which have become increasingly politicized. Secondly, I approach the topic in a semi-autobiographical and quasi-ethnographic way. I was immersed in the music industry from 1992 to 2008 both as an industry management *insider* and latterly as a more objective doctoral researcher. During these years I also gained experience of the film and theatre industries, and subsequently worked in the publishing industry from 2008 to 2012, which allowed me to build on my observations as part of a wider commentary on the cultural industries. I define *cultural industries* as any activity involving the production of content which represents, interprets, informs or entertains. 'Content' is of course a troublesome and potentially misleading metaphor. Ironically, it appeared in common new-media usage just at the time when paper, celluloid, vinyl and polycarbonate 'containers' of creative output began to be rendered obsolete by digital distribution. This illustrates the urgent strategic need of rights-holders to make the ontological claim that content can exist independently from its format or from its container. Notwithstanding the opposing arguments to this claim, content is still quite a useful term in encompassing texts, sounds and images presented or exchanged through a variety of media. Of course, 3D printing-copying-manufacturing technology may yet expand further the concept of what being in the 'content business' means, but that is a subject for another book.

A contextual focus on the strategies and the commercial organization of the production of texts, sounds and images necessarily draws in the inextricably related history and future of intellectual property rights, laws and policies. The discursive tactics employed by all sides of the still raging ideological battle over intellectual property in the digital age form a major theme of the book. The winners and losers are being defined through an ongoing storytelling contest which is rich in metaphor and competing social identities. Influence and power within the cultural industries still depend as much, if not more, upon the preservation of certain traditional societal roles and identities as they depend upon the intellectual property law which underpins their influence and power. Challenging this dominant power, new tales abound which portray traditional stakeholders with their heads in the sand, in stubborn and complacent denial of the inevitable consequences of convergence in media and technology, or as dinosaurs incapable of adapting to a changing environment. The old-world 'heroes' of inventors, talent-discoverers, patrons, established editors, journalists and other cultural interpreters and guardians are often now portrayed as corporatized, privileged and elitist, abusing intellectual property law to restrict the generation of, and access to, cultural capital. Using a discourse of social justice, an alternative new millennium 'wiki' world is proclaimed by those bent on disrupting the old rules and alliances. Between these constructs of new and old worlds, the technologists and the free-content disseminators may be alternately labelled as cynical pirates or as visionary democratizers, depending on the political agenda or strategic context in which they are being described.

The scope of the themes is therefore broad, but the lasting value of the book, I hope, is that it sheds light on some less obvious, less analysed and less well-understood obstacles to industrial and organizational change. In particular, it is concerned with the less discernible socio-political properties of language that construct robust narratives and support powerful identities which can either promote, but more often block opportunities for social and industrial change. Whilst I don't necessarily equate 'change' with 'progress', I do believe that obstacles to change rarely produce happy endings in the long run.

The roots of the book are in my doctoral research, which comprised a series of dialogues with influential stakeholders who are illustrative of varied interests in the economic future of music. The research itself was born of the desire to make sense of my organizational life and of my executive responsibilities inside a music company in 2006. This was a time when the recorded music industry was already in its seventh successive year of global decline towards being, in 2013, less than half the size it was in 1999 on an inflation-adjusted basis. From experts and amateurs alike, there had already been some good incisive journalism and persuasive diagnostic and prognostic theorizing on the fate of the music industry in the new millennium. Yet my curiosity was not entirely satisfied with the simplicity of the dominant post-Napster critiques citing the adaptive failure of record companies and their fat-cat managers. I felt that the prevailing interpretations were compelling mostly because they were plausible, easy to understand, and above all because they conformed to the overwhelming narrative of the digital revolution. Whilst I didn't disagree with many of their interpretations and insights, I found that their significance and usefulness were overstated, and often melodramatic. Sure, they were rich with metaphor which resonated with the progressive and emancipatory euphoria of the digital age and the inexorable technological march of the networked, sharing, collaborative society; but they were typical of revolutionary optimism in the afterglow of the overthrow of an established order: heavy with promise of something better, but light on the specifics of how 'better' might be achieved, or even on what it really looked like. By 2009, when I completed the initial research, and I had left the music industry for the publishing industry, there was more than a whiff of anti-climax in the air. There was also an anxiety that a revolution may never be more than a process between two states of power-oppression and that some forms of power-oppression are simply more recognizable and more explicitly organized than others.

The initial research was corporately sponsored by my employer (the ailing EMI), and the ostensible academic focus was located within the well-established fields of strategy and organizational change. It thus logically took place under the institutional authority of a business school. However, as the research proceeded it became clear to me that whilst the categories of social-science are administratively useful when co-locating individuals and texts with similar research specialisms, they also introduce artificial boundaries to the contemplation of complex economic and social phenomena. I sought answers to questions of technological evolution, of economic and aesthetic value-perception, and of behaviour, both individual and organizational. This curiosity required me to draw from many disciplines beyond those generally thought of as the primary domain of business schools. These include sociology, economics, linguistics, history, politics, psychology and law.

On this journey I was, for a while, intoxicated by the revelatory power of critical theory to highlight the dangers of society's prioritization of the *means* of wealth production and its neglect of the *ends*. I became disturbed by the tricky alliance between power and knowledge, and I had some of my taken-for-granted views of the world de-stabilized by the late 20th-century intellectual preoccupation with any theories prefixed by 'post-': e.g. post-structuralist, post-modernist, post-ideological, post-materialist. These things began to mean far more to me in my 40s than they did when I was 21. Flirting with heavyweight intellectual texts, I felt rather clever and insightful for a time – at least until it struck me how unlikely it was that an industry executive-turned-scholar would add anything radically new to the canon of literature in these rarefied fields.

Where I could contribute something original, however, was in linking some of the largely abstract socio-economic and political themes together in a new context, and in particular using examples from my experience as a cultural industries insider. I believe that *context* is vital in the production and the dissemination of knowledge. I share a post-modernist suspicion that the grand promises of social progress through improvements in universally applicable science and rationality are overly optimistic; that the coherence and autonomy of the human race is often illusory, as is much scientific (or at least socio-scientific) knowledge which makes ambitious claims to be valid beyond the local situation or context whence it was constructed. I therefore hope that a contextual and semi-autobiographical approach to the book will bring illustrative colour, depth and authenticity, providing I can be critically reflexive towards my own biases. In this way, I aim to make esoteric concepts and theories more accessible and industrially relevant, not just for scholars, but also for executives, managers, politicians or any other stakeholder with an interest in the cultural industries.

Discourse and power

The cultural industries encompass the commercial prospects of corporations with interests in technology, communications, media and content, and the intellectual property which protects those interests. This means that this book necessarily deals with some big themes of language and power in society. I should therefore explain from the outset why I believe that a critical language-based approach has much to offer those who want a deeper understanding of the fate of the cultural industries. I deem it appropriate to study these themes and rules as socially constructed because the successes and failures of organizations in the digital age are heavily dependent on the effectiveness of the discursive resources which are available to support them. In using the term 'discursive resources', I mean all of the various aspects of the domain of discourse, including: (1) the quality and balance of dialogic interactions; (2) narratives and stories; (3) rhetoric; and (4) tropes such as metaphor and irony which can alternately reinforce or disrupt prevailing paradigms. The debate over intellectual property reform in the digital age, a debate which many see as central to the fate of the cultural industries, serves to illustrate why this is so important. Since 2000, there have been several UK government-commissioned reviews of the economic effectiveness of intellectual property law. All have concluded that policy-making tends to be more influenced by stakeholder rhetoric, rather than by rigorous and measurable evidence-based research. This shouldn't be surprising; the topic does not readily lend itself to rigorous and robust empirical investigation with precise and certain outcomes. It therefore relies on untested hypothesizing and the projection of scenarios. As a consequence, plausibility, retrospective sense-making, and eloquent appeals to common sense prove to be more accessible and digestible sources of decision-making. Plausibility trumps precision every time. The problem is that common sense spawns two plausible, but fundamentally different, positions in this debate: intellectual property rights and laws should be loosened, and they should be tightened.

Illustrations of the power of stories and of rhetoric may come as no great revelation, so why am I so concerned about a 'critical' approach which implies a necessary unmasking of scarcely discernible but dominant or restrictive ways of seeing the world? It is because I fear that we often delude ourselves about our ability to stand back and view the world critically. When I am *not* too tired or too busy to think and reflect, I am conscious of being subject to the rhetoric and sophistry of political spin, brand marketing, and other well-packaged media interpretations of how we should understand and engage with the world. But a busy and distracted life means that I *am* often too stretched to sustain persistent critical thought. I cannot unpick every cultural symbol, whether linguistic, physical or ritual, in order to reassure myself that I have not unwittingly become enthralled by it, and that my identity has not in some way become problematically bound to it. Yet whilst I acknowledge the presence and power of symbols, my education and my professional success nevertheless lead me to believe that I remain the enlightened master, rather than the subject, of my language and interpretations. I find it easier to convince, or delude, myself, asymmetrically, that I can influence others by the language I use, than to accept the full extent to which I myself am subject to, and a product of, certain prevailing language constructs.

The presumption of objective consensus around many of these linguistic constructs became so well established in the 20th century, or earlier, that they are taken for granted as self-evidently optimal ways of organizing society: e.g. democracy, universal human rights, freedom of expression, the progressive societal benefits of scientific research and technological innovation, the independent rule of law, capitalism, copyright, the limited-liability corporation, managerialism, double-entry accounting measures (debits must equal credits), freedom of trade and of markets, and the pursuit of growth in national productivity (GDP). These concepts form part of the powerful meta-constructs, or ideologies, which shape the way society is organized, *and* they shape the values and the conversational mainstream of its citizens. Each of them has broad socially embedded and reinforcing vocabulary and tropes that constitute a *discourse*, or interpretative repertoire for how to live, how to think and how to 'be' in the world. They pass for 'truth' or common sense and make it difficult to see the world in alternative ways.

Social scientists, historians and intellectuals started questioning these constructs from the moment they were given names. Sometimes they challenge the ideological principle itself, as is the case with capitalism or neo-liberalism. In other cases the critical lens is focused more on the obsolete or corrupted institutional practices which are carried out in the name of the higher ideologies. One example would be the concern that the relentless pursuit of profit and productivity growth for its own sake has unintended consequences which conflict with other sacred principles; e.g. human rights, especially the human rights of the citizens of 'less-developed' nations not to be 'globalized' and to organize themselves more sustainably using broader measures of wellbeing; or the rights of future generations not to be crippled with the debts of their ancestors, nor with a planet that cannot support them. Another example, more closely related to the theme of this book, would be the assertion that the rule of *law* has been replaced by the rule of *lawyers*, illustrated by an estimated \$20 billion spent on costs to defend legal patents in the smartphone/tablet industry in recent years. The relationship between language, knowledge, power and politics is also richly illustrated by the rhetorical battle over the question of whether copyright laws are fit for purpose in a digital age. The 21st-century challenges to the concept of 'natural' intellectual property rights threaten to weaken the mainstay of capitalism in the highly competitive knowledge economies, hence the vigorous counter-arguments.

For many years, and escalating notably since 2007, there have been sustained discursive (and physical) protests against the hegemony and destructive imbalances caused by the ideology of capitalism, multi-national corporations and global financial systems. As powerfully apocalyptic as those protests continue to be, and as precarious as the western financial systems and debt projections are, nothing has (yet) de-stabilized the dominant capitalistic and managerialist culture of organizations. Balanced scorecards attempt to broaden the key measurements of organizational performance with reference to corporate strategy and mission, but the financial language of capitalism is still regarded as the most robust basis for measurement: risk, reward, returns on investment, property rights and their protection, tangible and intangible assets, shareholder value and success defined narrowly by profit growth expectations. 'The bottom-line' is not merely the net sum of a column of revenues and costs defined by accounting standards. It has become metaphorically synonymous with the most important and definitive final measure or outcome, not only of a commercial enterprise, but of any analysis or discussion. The rise of the language of 'corporate social responsibility' may have reined in some of the more rampant corporate exploitation of the 20th century, but *the corporation* is still a product of its legally obliging structure: a hard rational profit-generating institution. Whatever the worthy moral intent of some of its managers, for the most part the corporation squirms uncomfortably with the idea of social responsibilities, or at least those responsibilities which cannot readily be measured and justified as ultimately enhancing shareholder value through corporate brand reputation.

A few politicians and a growing proportion of the concerned public have joined the critically aware ranks of the discontented, especially following the economic downturn of western nations since 2007. Yet despite this uprising against the social and global inequities of 20th-century institutions, the dominant practices of those institutions and ideologies are still not regarded as especially contentious or problematic within the everyday organizational life of the majority of corporations. It is unfashionable to say it, and consultants may protest, but in spite of huge and continuing change in products, in corporate winners and losers, and in the make-up of 'value chains', little has changed in the fundamentals of economic measurement and management practice in the past 15 years, at least within the cultural industries.

The cultural industries are not necessarily the primary concern of the malcontents and the political activists, as compared with, say, financial services, bio-techs or the extractive industries. Nevertheless, the slowness of corporate organizations and of state policy-makers to respond to 21st-century malaise is still of great relevance to the question of how cultural production should be organized. Anti-capitalist discourse includes the assertion that 'it is easier to imagine the end of the world than the end of capitalism'. The obscure authorship of the phrase suggests it has already gained proverbial status, and implies that these are desperate times indeed. Despite broadly felt anxieties that the world does not (at least not yet) have self-sustaining economic solutions to match the aspirations of its resource-hungry inhabitants, belief in the future continues to rest upon a blind faith that the old economic patterns will cyclically return, *if* companies and governments can hunker down for long enough, and *if* their knowledge economies can maintain a competitive edge in *innovation* in both technology and marketing.

I highlight the word *innovation* here because it is innovation which brings us back to the theme of intellectual property. For businessmen and politicians alike, innovation has become the uncontested economic silver bullet for corporate success and national economic growth. But innovation and intellectual property protection have a complex relationship. On the one hand they are seen as economically interdependent: innovation may not occur on any meaningful scale without the incentive provided by the legal protection of ideas. On the other hand, patents, copyrights and the processes and behaviours that accompany them often stand in the way of the fluid and rapid exchanges necessary for the development of new ideas and solutions. Furthermore, the private ownership of knowledge means that the social benefits are often limited to only those who can afford to pay. Slavoj Zizek, a well-known promoter of the aforementioned proverb, regards society's failure to reform intellectual property laws in the 21st century as one of the four horsemen of his predicted apocalypse. Though pharmaceuticals and bio-genetics may steal the headlines in his vision, all attempts to enclose the commons of the mind are seen as morally unsound. Contemplation of the future of the cultural industries can thus ignore neither the intensifying challenges to intellectual property protection nor, by extension, the larger societal malaise with capitalism.

The cultural industries, as we currently recognize them economically, would be decimated without any of the protection of the private property rights enshrined by capitalism. I make this comment merely as an observation, not as a defensive argument for the status quo in statutory rights protection. The large-scale industrialization of culture was only made possible in the UK through the statutory recognition of authors' output via economic artifice over 300 years ago in the Statute of Anne (1710), well before the industrial revolution. Copyright historians reveal that the original intent of copyright law was as a short-term (14-year) state-granted monopoly privilege, a kind of tax, creating an artificial scarcity for non-rivalrous goods, i.e. replicable goods that one person can enjoy without limiting the enjoyment by others. Anything which stood in the way of the free flow of science and knowledge was anathema to the Enlightenment. A monopoly on ideas was conceded then as a necessary evil to stimulate behaviour which would promote further knowledge and learning, to the benefit of a progressive commonwealth. The temporary granting of exclusivity to an idea, the 'fugitive fermentation of an individual brain' was not a natural right, but a 'gift of social law', 'done according to the will and convenience of the society, without claim or complaint from anybody'. This was the view expressed by Thomas Jefferson in 1813. It was only later, being subsumed into the industrial revolution, that this fiscal privilege became discursively re-constructed by industrial capitalists as a 'natural' 'private' 'property' 'right', and its term was extended accordingly (potentially more than tenfold, now to the author's life, plus 70 years). Prior to being again disrupted in the digital age, the capitalistic rationality of copyright had become so culturally embedded in the 20th century that it was regarded as indisputable common sense. Yet in its earliest statutory 18th-century incarnation, copyright law was more in tune with the many 21st-century calls for its reform, in particular the relaxing of rights in order to promote knowledge-sharing, innovation, economic growth, a more level commercial playing field and a fairer, more open society. There is now a large gap between the rules of law and the new social norms shaped by technology. As a result, the law is rapidly losing the support of the general public, and of business.

Arguments about copyright in the digital age have become as entrenched as they are polarized. Authors, artists, producers and content-owners argue that they have a right to fair compensation. Their argument is rhetorically defensive and can be summarized as follows: 'we are the biggest exporters and drivers of economic growth; we are defenceless against mass-scale infringement, directly or indirectly supported by complicit new media players; therefore we need more protection'. The opposing view challenges traditional content-owner arguments: 'you abuse copyright to conserve monopoly power and sustain outdated business models; in doing so, you stand in the way of innovation, creativity, progress, economic growth and social justice'.

Economic growth prospects may depend on how creatively these conflicting ideologies are reconciled, but on close scrutiny these arguments do not need to be so polarized. Demanding an either/or choice is itself a melodramatic discursive tactic designed for political ends. The fear and mistrust of opposed parties that government will be swayed too far in one direction mean that the agents of each side of the argument feel duty-bound to their stakeholders to make a big noise. Each invokes their own melodramatic visions of the societal damage which will ensue from certain courses of action. However, there *are* moderate compromise policy solutions available, even if they are complex and tiresome to legislate. Some progress is being made, but at a snail's pace. This is evidenced by the fact that the main anti-infringement provision of the UK Digital Economy Act (2010) will not be implemented in the foreseeable future due to the wranglings of the various affected parties. Rising above these inter-stakeholder altercations, it is important to keep in mind that the highly evolved, subtle and often indiscernible cultural 'value chain' processes of patronage, curation, dissemination and the facilitation of consumption of texts, sounds and images are all as important now as they ever were, even though they have had to adapt to new technologies and to powerful new market entrants. Within industries which only really exist because of technologies (e.g. printing, recording, broadcasting and digitization) and which have long suffered cycles of death and re-birth through new technologies, the winners will always be the innovators, in consumer marketing as much as in technology. The moderation of copyright law, whether by loosening or by tightening, may moderate the pace of change to advantage some stakeholders over others, but it will have less aggregate impact in the long term than much of the fear-mongering rhetoric would have us believe.

Twenty-first century government policy is, perhaps rightly, obsessed with the national capacity to innovate technologically and creatively for international competitive advantage across all industries. This means that intellectual property policy is preoccupied with the question of whether current laws stand in the way of innovation, and whether constraint in innovation means constraint in economic growth. Non-governmental initiatives such as Creative Commons and the Open Access movement have already made big strides forward in changing attitudes to the unfairness of excessive rights protection, but government initiatives have been fraught with difficulties. In principle, I am in favour of the reform of copyright law, but I imagine that the very long lead-time which would be required to make any really meaningful transformation mutually acceptable to all parties will be politically unachievable. The road is too long and uncertain, and the leap of faith too great for many stakeholders. On a more optimistic note, one of the observations of this book is that the statutory reform agenda is a slow and distracting sideshow to the main obstacle to innovation.

That obstacle is the difficulty faced by content, media and technology stakeholders to collaborate effectively with one another. This is largely based on a fear of losing a precious place in the value chain. Technology and new sources of competition make it difficult to experiment for fear of setting an irreversible precedent. At best, interactions are inefficient; at worst, major opportunities are lost. The absence of compelling consumer offerings for music via mobile telephony is an example of such missed opportunities. The diagnosis of such commercial discord cannot simply be put down to problems with copyright permissions clearance, nor to the fear and mistrust of collaborating with one's potential competitors and disintermediators. The cause is more embedded in conflicting language constructs, and the quite deep-rooted social roles, responsibilities and worldviews they support. I believe that resources which address how these behavioural obstacles to collaborative innovation might be overcome would have a more fruitful outcome than merely tinkering with the impenetrable catch-all spider's web of copyright law.

The title of this book alludes to the character of some of the roles and identities of the stakeholders referred to above. These can become so rigid and entrenched that we forget that they may be merely provisional, negotiated and mutually convenient ways of seeing the world. Instead they pass for a more permanent reality or truth which then shuts down certain courses of action. The channels for invention and innovation seem to me to be no less prolific and socially accessible today than they have been historically. Legal modifications to intellectual property protection may allow some innovations to proceed more freely, and may alter the relative power, influence and size of commercial participants in the cultural industries value chain. But such statutory modifications will not unlock the larger psychosocial obstacles which stand in the way of a leap of faith towards more interesting strategic and commercial collaboration. Such collaboration would ultimately be a much greater determinant of value and growth in the cultural industries. This book aims to divert attention away from the old master-narrative of cultural production, i.e. the destructive feud between old and new stakeholders over intellectual property reform, and to present obstacles to innovation in a new light which may diffuse conflict.

I want to be clear from the outset that this is the kind of book which raises more questions than it offers answers. The book does contain some views about the future, but its primary value is not in offering easy prescriptive fixes for the cultural industries. It is more about provoking its readers into new critical ways of thinking about the changing qualities of dialogue and collaboration in a creatively and intellectually 'post-private' world; a world which is still stuck in the muddy obsolescence of the 20th-century constructs of cultural intermediation.

Structure of the book

The book is something of a hybrid, being motivated by an industry insider wanting to combine scholarly rigour with business relevance, and a touch of autobiographical informality. It is probably helpful to state upfront that I started the book with three exploratory goals. The first was to reflect upon the record industry's response to the phenomenon of file-sharing and the demise of recordings as physical artefacts, and upon whether we could have done things differently. The second was to explore the extent to which discourse analysis can contribute to the better understanding and practice of strategic management. This has been the subject of much academic theorizing, but is rarely (if ever) originated and presented from an industry insider's perspective. Thirdly, I felt compelled to link my insights and experiences to the ongoing deadlock in copyright reform, and to show how opposed constituencies impede resolution through their different self-interested constructions of the world of cultural-industrial production. The political-industrial quagmire on this topic over the past 15 years is not only enormously wasteful of resources, but will in the longer term more likely damage than benefit the cultural industries of the developed world. If I can make even a small contribution to moving the debate forward, the book will have been worthwhile.

The diversity of these goals does present something of an editorial challenge, and breaks the mould of the traditional scholarly book format. I have thought carefully about how to present the material and have split the content into four parts. Each part has a distinct style and character appropriate to its content, and may appeal to different readers in different ways.

Part I, *My Version of Events*, presents a particular view of recorded music history, and of the industrial and organizational master-narrative upon which its identity, cultural influence and economic power have been based. Chapter 1 begins with some personal anecdotal material from the heady optimism at the dawn of the new millennium. Chapter 2 then tries to make some sense of those years by locating them in the story of recorded music industry history from its origins in the late 19th century to the present day, tracing certain threads such as the characterization of the pioneer-protagonists in the story and their changing relationship with technology. Although this history aspires to objectivity, it is inevitably coloured by my own experiences. I invite readers to reflect and draw their own conclusions about the extent of any residual bias I may have towards the industry.

Part II, *Stakeholder Voices*, is the empirical heart of the book. Drawing from the source material of my conversations with senior and influential stakeholders in the music industry, old and new, it presents a range of industrial views, using a high proportion of their own transcribed words. Chapter 3 lays out the traditional value chain of recorded music and ancillary business models, and then tracks the shifts in perceptions of value through the participants' own voices. It includes the impact of the digital commoditization of music and the apparent diminution in the value attributed by consumers to high quality recorded sound and physical packaging, in favour of convenience and services which enable discovery, sharing and personalization. It explores the value propositions of competing stakeholders, who have an interest in devaluing some product conceptions of music relative to their own consumer offerings, such as mobile phone companies and social-networking sites. Chapter 4 introduces the concept of a custodial dilemma which is at the heart of most of the conflict in the cultural industries: how to balance the protection of, on the one hand, a successful and centuries-old privatized system of cultural intermediation, and on the other, the civil rights to access and to generate cultural capital in new ways, made possible by new media and technologies which show scant respect for the old institutions. Chapter 5 concludes Part II of the book with some hindsight interpretations of what happened in the music industry in the previous ten years. Illustrations are given of how perceptions and openness to organizational change can be desensitized by the kind of clear and public commitment to strategic choices and visions which industry actors feel compelled to make.

Part III, *A Storytelling Contest*, focuses critically on the language utilized in the discussions of the music industry's past, present and future, with a particular emphasis on the various narratives at play. For those not familiar with the field of discourse, Chapter 6 gives an overview, explaining the various domains of discourse, and the epistemological claims of discourse analysis. Chapter 7 introduces the concept of *strategy as storytelling*, using examples from the music industry. It also considers the role of strategy as a sense-making process for executives. Chapter 8 uses the research texts to highlight the most recurrent and relevant discursive objects, for example: music, the consumer, the record company and technology. These objects are shown to have quite distinct and sometimes competing linguistic constructions. Chapter 9 then reveals a series of narratives, each defined by their protagonist, including the patron, the curator (in various guises) and the inventor. I broadly divide these tales into two groups based on the worldview which they adopt. *Tin Pan* tales are those which promote a more capitalistic view of the music industry, whereas *Wiki* tales are products of a web 2.0 cultural ideology. Chapter 10 takes one of the tales, the Inventor's tale, to explore a fundamental difference in the conception of technology, and contemplates the consequences this difference has for future strategic possibilities for music organizations now that the industry no longer controls its own technology. Chapter 11 concludes Part III of the book with a critical reflection on the themes of power and ideology at play, and why winning the storytelling contest is so significant.

Part IV, *The Pirate's Tale: Copyright Reform, and the Future*, concludes the book with what is thought to be the most important question which currently faces the music industry, and indeed the whole of the cultural economy: the reform of copyright. Chapter 12 picks up on a traditional counter-narrative which was notably absent from the stakeholder conversations, namely the pirate's tale or, more precisely, the important variations in the narratives which interpret the motives of those who ignore the law. It charts the origins of the construction of the pirate in the modern age and how it has been transformed in the last 15 years to suit alternative ideologies and political agendas. It concludes that the popularity and moral ambiguity of pirate narratives play an important role in questioning the relationship between the individual and the state in matters of creativity, knowledge-sharing and trade. Chapter 13 introduces the newer counter-discourse of cultural conservation and of the public domain. It suggests that this discourse lacks the overarching framework and emotional connection of a master-narrative. Being more defined by its various antagonists than by a clearly drawn protagonist, it is consequently falling short of the worthy aims of its proponents. Chapter 14 builds on the counter-discourse of chapters 12 and 13 and presents a brief history of the intellectual and political narratives which have shaped copyright law, and the associated conversations about the legitimacy of the claims of authorship and protection. It reflects on whether or not we are living in a new age of cultural production, and weighs up the likelihood and desirability of copyright being reformed in the second decade of the new millennium, with particular reference to the UK Hargreaves Review (2011) and Maria Pallante's 2013 call for 'the next great copyright act' in the US. Chapter 15 concludes with some reflections about the changing behaviours in cultural consumption, some utopian and dystopian visions of the future of cultural production, and encourages more public engagement with the topic of copyright.

Contribution to fields of study

Where does this book fit within the literature of the fields into which it steps, and who should read it?

It aspires to contribute to a number of different fields which are best described with reference to just a few of the texts I have found most enlightening in my journey. These can be arranged into three groups.

The first contribution is as a rare record of first-hand experiences and conversations which are relevant to those who are interested in piecing together a recent history of the music business. Amongst the superabundance of writing about the music industry, there are three texts I want to highlight because they helped me to locate the relevance of my findings, and to create some new insights by connecting certain themes. These texts are particularly original and relevant to a genealogy of recorded music discourse. Roland Gelatt's *The Fabulous Phonograph* (1955), now out of print, is a meticulously researched early history which has the benefit of not being retrospectively analytical from a digital perspective. Mark Katz's *Capturing Sound* (2004) is a wonderful demonstration of how recording technology itself has agency in the music industry narrative, going way beyond being just a neutral mechanism for capturing and preserving artistic expression, and making its own claims as a dynamic and independent source of creativity and value. Finally, Jonathan Sterne's *MP3: The Meaning of a Format* (2012) has been especially useful in helping me connect my first-hand experiences of PolyGram and Philips in the 1990s to the parallel and subversive story of MP3 in order to piece together precisely when and how the divergence of industrial goals crystallized competing conceptions of the cultural production process, and broke the century-long corporate symbiosis of music and technology.

The second contribution is to the literature of the cultural industries more broadly, and in particular the dynamics of legitimacy, authority, power and protection in the processes of cultural production. Although the book is empirically centred on the recorded music industry, its themes are common to most areas of the cultural industries. David Hesmondhalgh's *The Cultural Industries* (2013) has been a valuable comprehensive work in helping me understand some of these commonalities and connections, and hopefully in return offers some new 'insider' insights for the ongoing study of what has changed and what remains the same in cultural production. Martin Parker's wonderfully crafted paper, *Pirates, Merchants and Anarchists* (2009), introduced me to a new perspective on cultural representations of alternative forms of organizing and to an important but complex master-narrative which was notably absent from my primary research sources. An interest in the pirate's tale also led me to Adrian Johns' *Piracy* (2009), which has provided a thorough and enlightening source of the history of the social construction intellectual property and its infringement. Whereas Johns' focus is on sociological dimensions, I found that the contested legal history of copyright is eloquently demonstrated by the series of essays entitled *Privilege and Property* (2010), edited by Ronan Deazley, Martin Kretschmer and Lionel Bently. No study of the new narratives of cultural production and the law could omit the works of Lawrence Lessig and of James Boyle, who are amongst the most articulate, rigorous and consistent intellects of the cultural commons movement. Finally, Jessica Reyman's *The Rhetoric of Intellectual Property* (2010) is worthy of special mention as a relevant and important scholarly work, which opens up lines of inquiry which this book aims to address. I concur with her view that the various groups which portray the inequities and obsolescence of copyright in the digital age have yet to find a compelling heroic protagonist for their user-generating cultural conservancy counter-narrative. Beyond these contemporary texts, the book draws from a post-modern well of scholarship concerned with the diminishing influence of the linear literary mind, and the dubious legitimacy of elite cultural originators. These views emerged more than half a century ago with colourful disrupters such as Marshall McLuhan, Roland Barthes and Michel Foucault, but seem as relevant as ever when considering the proliferation of texts and of the claims of authority, influence and the need for legal protection in the second decade of the new millennium.

My own contribution to this second broad and valuable field of scholarship is in helping to make it more relevant to a non-academic readership by locating within it my experiences of the collaborative obstacles and strategically entrenched positions of the cultural industries, and in doing so, presenting issues in new ways which are more visible and discussible.

The third contribution is to the field of organizational discourse, strategy and change. The contribution is *not* a theoretical one, but more of an unusual experiment in how industrial management practitioners might valuably engage with academic theory relating to discourse analysis. The field is informally articulated through the International Centre for Research in Organizational Discourse, Strategy and Change which connects dozens of eminent academics from around the globe. It was here that I had an epiphany in 2001. Through the works of scholars such as Mats Alvesson, Hugh Wilmott, Cliff Oswick, and Tom Keenoy, to name just a few, I discovered that the behavioural, organizational and industrial problems with which I was most concerned could not be satisfactorily addressed without a better understanding of the relationship between language and power, and in particular, a critical approach to the analysis of management and industrial discourse. The original research problem was shaped by Norman Fairclough's framework for critical discourse analysis, where a key area of focus is to identify social constituents who have an interest in particular social problems *not* being solved. Karl Weick's work on sense-making resonated profoundly with my experience of corporate strategy and all its idiosyncrasies; and my narrative approach draws from various authors who share an interest in organizational storytelling, such as Yiannis Gabriel, David Boje, David Barry and Michael Elmes. The highly intellectual profile of the organizational discourse field tends to lead to its publications being characterized by an interest in the abstract and epistemological dimensions of discourse. It therefore exists on an esoteric fringe of management and organizational science, where contextual, empirical projects with rich data sources are not common. Consequently, there seems to be reluctance within business schools to embrace organizational discourse within mainstream research and teaching. Although some discourse theorists may find some flaws in my pragmatic approach, I hope that this book, being data-rich at its empirical core, and being written by someone who is a management practitioner first and a scholar second, goes some way to stimulating a wider practitioner interest in the topic and making it more accessible to new audiences who would not otherwise venture into this fascinating field.

The book is ambitious in the breadth of its aims, and I acknowledge that its thematic and stylistic diversity makes certain demands upon the reader. I have, however, tried to preserve a kind of organic evolutionary quality to the content as it emerged, and hope that this will preserve the structural integrity of the book, and make it a rewarding read.

Ultimately, although the book derives most of its plausibility from analysis of music industry discourse, I hope that it will have something to offer any reader who is interested in the fascinating 21st-century shifts in the broader industrial processes of creation and dissemination of knowledge and culture.

Part I
My Version of Events

1
A Personal Perspective

This first part of the book is my narrative attempt to make sense of recorded music industry history. It is shaped by my subjective observations and experiences – as an active consumer since the mid-1970s, as a finance professional and industry insider from 1992 to 2007, and aspires towards objectivity through my efforts as a scholar since 2005. It does not claim to be a definitive interpretation, but simply a reference point for critical reflection in subsequent chapters. Before I begin, it is worth asking: what is the point of trying to make historic sense of the music industry?

I have two reasons.

The first reason is specific to the music business. The major record companies are generally considered to be amongst the first, and certainly the most talked about, industrial casualties of the internet age. Everyone seems to have a view on what happened to the music industry, especially anyone who takes an interest in music, or in the creative industries, or in technology and media innovation, or in organizational strategy and change management, or in intellectual property law and cultural-economic policy. That adds up to a lot of people, and most have formed their own theories on why it happened and what, if anything, might or should happen next. Prevailing popular discourse constitutes the internet and new social media as vehicles for a liberating and field-levelling revolution. The discourse is antagonistic, even belligerent, with the implication that the repressive bastions of the old guard need to be overthrown. But whilst the old guard have been wounded, and some even died, those who remain soldier on in a battlefield whose lines have been re-drawn. Some are bigger and stronger than before. There remains some antagonism towards big corporations that still play the role of cultural intermediary and use their shield of intellectual property protection to hold on to power. Wonderful new devices and channels provide more consumer choice and empowerment than ever before. It is claimed that budding musicians can be seen and heard more easily than ever before. Yet for some reason there remains a sense of injustice about where the power and influence lies, and how the spoils are divided.

I understand the disappointment and frustration at the endless war of words between those who want a free flow of musical expression and discovery and those who seek economic reward and recognition for their contribution to cultural production. But the war is not a simple two-way contest. Those who claim that rights-owners are obstacles to innovation and cultural democracy do not speak with one clear voice. They include the highly vocal and articulate cyber-utopians as well as the more muffled ventriloquial corporate voices of new media service providers and device manufacturers. This ventriloquism derives from the recognition by companies such as Google, Facebook, Apple and Amazon, amongst many others, that their strategic goals may be as misaligned with the dreams of the cyber-utopians as with those of the music and film companies. They rely on others, including users who are addicted to their products and services, to publicly lobby for less restrictive copyright. For now, there is a sense that not much has changed at all and that the real revolution has not yet happened. Identifying the likelihood or desirability of a further revolution in copyright protection is the first reason I am trying to make sense of the music industry.

The second reason is because the music industry, constructed as a valuable national domestic and export asset which has fallen victim to the unregulated wiki wild west, tends to be stealing the limelight in the wider political debate around the rights, controls and protections attaching to expression and innovation in education, culture, science and technology. There is thus a real danger that innovation in the 21st century, and in particular the future of the internet, and the way it is regulated and policed, will be disproportionately influenced by discourse which has a nostalgic sympathy for the music industry. Reconciling the competing interests in intellectual property policy reform is a fundamental social dilemma, just like many other political balancing acts which tease and torture societies. In this context, the second reason for beginning the book with some history is to establish a balanced long-term perspective which encourages critical reflection on the dilemma, and will hopefully promote more open-minded dialogue.

New millennium dawn, or dusk?

In February 2000 I returned to the UK after seven years working overseas for the world's largest music company, Universal Music Group. My original assignment had been to New York, then Madrid and ultimately Los Angeles. Still employed by Universal, I returned to the company's international headquarters in London, where I was asked to project-manage the development of an embryonic and innovative concept called Voxstar. The project aimed to capture the potential of new media and technology in order to transform the consumer experience of engaging with music and artists. My return coincided with the peak of the dot-com boom, and, as it turned out, the financial peak of the global recorded music market, though no one knew these things at that time. London was buzzing with the anticipation of further extraordinary growth yet to come. It was a particularly exciting time to be in a record company: full of possibilities, bright sparks, blue ocean visions, and beautifully constructed PowerPoint slides.

Voxstar was designed to create a new technology platform for music, not just for Universal's content, but open to all artists; and not just for the paid download of recorded music, but offering a broad experience of content and community, news and chat, access to artists, games and competitions, streamed music, and live webcasts. All this would be provided interactively via new digital technology, channels and devices as they emerged. Moving away from the old business model of music to be sold and owned, there were new experiential values to be exploited and monetized, at least according to the new-age dot-com consultants supported by the neat precision and predictability of their spread-sheet models. Choice, convenience, personalization, empowerment, flexibility, discovery, immediacy, exclusivity, community and mobility were the determinants of value for the 21st-century digital consumer; or so we believed. As a logical consequence of serving these needs, revenue would flow from a number of new channels including subscriptions, 'freemium' products and services, advertising, sponsorship, licensing, artist-branding, merchandizing, ticket-commissions and more. The main risks, as they appeared then, were not that we were strategically misguided, but that others would beat us to market, or that we would be blind-sided by a 'killer-app' which eclipsed our own innovations.

Universal committed over £15 million of development money to Voxstar but we sought third-party funding, not only to mitigate the risk, but also to demonstrate the independence of the new venture from Universal's traditional business. The new millennium investment environment was intoxicating. Despite growing evidence of an already bursting dot-com bubble, in late April 2000 the major telecommunications operators still had the confidence to bet their futures in the largest ever UK auction. They bid a jaw-dropping and ultimately crippling £22.5 billion for licences to operate the third generation (3G) spectrum which they believed provided the platform to enable phones to take over from computers as the gateway to broad and 'rich' media consumption. Music delivered by mobile phone was very much at the forefront of these hopes.

In that climate, demonstrably prudent behaviour was symptomatic of lacking vision, and scepticism was for spectators and losers. It therefore seemed not at all absurd to find myself sitting in front of the world's leading investment banks, inviting them to commit £50 million for a 20% stake in something unproven and yet to be built, simply based on the expected future value of market-leading music rights exploited via new media channels. Only a few months later, our fate seemed secured when Voxstar was designated as the music partner for Vizzavi, the mobile internet portal recently constituted as a billion-euro joint venture between Vodafone and Vivendi. We stepped up our 'burn rate', or negative cash-flow as it should have been more soberly described. Many people deemed burn rate to be indicative of the boldness of one's ambition, being confident that revenue streams would emerge if the online user 'traffic' was 'sticky' enough. 'Build it, and they will come' was the mantra.

Sadly, Voxstar, Vizzavi, and 3G services in general were all ahead of their time, and naïve in their ambition. Aspirations for a brave, open and collaborative new world were misplaced. Despite being fully staffed and ready for launch by the spring of 2001, Voxstar was prevented from going public by a lack of confidence and consensus regarding competing digital strategic priorities amongst the various stakeholder decision-makers in London, New York, Los Angeles and Paris. In a tense and awkward meeting overlooking the Arc de Triomphe, Vivendi-Universal CEO Jean-Marie Messier and Universal Music CEO Edgar Bronfman Jr. instructed us to re-purpose Voxstar's leading edge technology to support other internal projects including the recently acquired MP3.com. The majority of the newly hired team was let go, along with their valuable new-media skills and their fresh outlook. Vizzavi suffered a similar fate, disappearing one year later. MP3.com limped on for two years but was dismantled and sold in 2003.

Another pair of early industry-led new-media innovation failures are worthy of mention. Pressplay, a partially collaborative online music subscription service which linked Universal with its competitor Sony, and with Microsoft and Yahoo, also closed in 2003. A rival music rental service called MusicNet, linking other music competitors EMI and Bertelsmann with AOL Time Warner and Real Networks, failed shortly afterwards. Both Pressplay and MusicNet failed for two fundamental reasons. First, they didn't license music content to each other, meaning that user access was limited to less than half the music which users would naturally be seeking. Second, neither service offered a satisfactory portable solution, effectively confining the enjoyment of the music to a user's computer. Lack of portability was not so much due to the often-cited technical obstacles as it was to an ultimately self-defeating timidity. Such corporate timidity was based on a pessimistic presumption that giving customers too much flexibility in their consumption would promote file-sharing. The thought process went as follows: 'if we give customers a service which is too good, they will abuse it', rather than a more confident and optimistic: 'if we give customers a great service which they really want, then they will happily pay for it and keep using it'. By not giving customers the benefit of the doubt, the music industry suffered a high price. In 2006, the then defunct services of Pressplay and MusicNet were ranked 9th in PC World's '25 Worst Tech Products of All Time', with the condemnation that the 'brain-dead features showed that the record companies still didn't get it'.[1] By then, the recorded music industry had given up any strategic hope or intent of retaining its mastery of technology in the 21st century in the same way it had controlled the technological means of production and consumption of music in the 20th century.

I mention Voxstar and these other heroic failures at the beginning of the book because I want to dilute, though not to dismiss entirely, the myth that record companies contributed to their own demise by burying their heads in the sand, refusing to see or to contemplate the impact of the internet on their business models. The myth continues that when they did finally begin to understand the opportunity, they were incapable of acting in anything other than a protective-defensive way, seeing only threats, not opportunities. This is a slightly misleading depiction in two ways. Firstly, it ignores the fact that some important open-minded and collaborative early attempts were made by record companies to innovate, albeit with potential collaborators who came with their own prejudices and obstacles. It should also be noted that all parties were operating within complex contractual, statutory and regulatory constraints. Secondly, the accusation of wilful myopia directed at record companies usually centres on their behaviour *after* 1999. By contrast, *my* diagnosis suggests that this hereditary affliction began much earlier, as I will demonstrate in the next chapter.

2

Innovation or Bust: A Short History of Recorded Music

It is not the purpose of this book to recount in detail the fascinating but complex twists and turns of the international recording industry since Edison triggered the start of the race in 1877 with his tinfoil cylinder. There are several books which document this very well, such as Roland Gelatt's early history, *The Fabulous Phonograph* (1955), or Mark Coleman's more modern journalistic account *From the Victrola to the MP3* (2005), to name just two.

Nevertheless, setting a historical context is essential for the critical reflections later in the book, so from these and other accounts, including personal experience, it is worth summarizing the music industry story at an unashamedly high level, with some relevant milestones in the evolution of recorded sound and its commercialization.

The first 100 years – fidelity and format dominance

The first 30 years of the recording industry were dominated by a format war between cylinders and discs. Cylinders were the domain of Edison, before being more commercially developed by the Columbia Phonograph Company, formed in 1888 by Alexander Graham Bell, amongst others. Discs were pioneered by Emile Berliner, and developed by the Berliner Gramophone Company (1895), and then by Berliner in collaboration with Eldridge Johnson through the Victor Talking Machine Company (1901) with its flagship disc player, the Victrola. Columbia and Victor became the two dominant competitors for the first two decades of industrial development.

Financially speaking, the sale of players, or 'hardware penetration' in late 20th-century speak, was perceived as being the big prize, rather than the sales of recordings themselves. Developing relationships with

musicians was a promotional strategy and Columbia and Victor com-
peted for talented performers to endorse their competing technologies.
By 1891, Columbia had recorded a catalogue of the popular military
band leader John Philip Sousa, along with a host of speaking recordings and vernacular songs. Competition for recordings of more serious
musical substance came from Victor through its Red Label, which was
supported by European classical singing stars such as Nellie Melba and
Enrico Caruso. Caruso's first recording was in 1902, and he went on
to become the first million-selling recording artist. Visual art also contributed to the competition between the formats. In 1899 the artist
Francis Barraud was commissioned by an interested party[1] to amend
his 1895 painting of his fox terrier, Nipper, listening to a phonograph.
He painted out the Edison cylinder phonograph, replacing it with
a gramophone disc player. The rights to the painting were acquired
and *His Master's Voice (HMV)* went on to become one of the most
recognizable trademarks of the 20th century.

In 1903 Columbia acknowledged the greater suitability of the disc format for musical reproduction and began manufacturing recordings in
disc format alongside its cylinder recordings. By 1912, Columbia discontinued cylinders altogether. Berliner's disc format had won, largely
due to Victor's marketing strategy of exclusively engaging celebrity
classical conductors, musicians and opera singers. Edison alone continued to produce cylinders, primarily exploiting them for his originally
intended purpose as a means of dictation, but the format was, by then,
moribund.

During the second decade of the 20th century the industry thrived,
and was led by Victor. By 1919, the US market for the industry's products
was worth $159 million.[2] In that year there were nearly 200 manufactur-
ers producing more than two million machines, and in 1921 production
of recordings exceeded 100 million units. By that time, two things had
become clear: first, that as the early patents expired, the business of
selling discs was at least as financially interesting as the business of
selling disc-playing machines; and second, that being closely in touch
with emerging cultural trends was as important as having exclusive
arrangements with established classical artists. The early leading role
of classical music in the shaping of the development of the industry
was soon overtaken by popular music, especially dance music, jazz,
blues and ragtime. Spotting the new sounds, and the dance trends, and
the emerging talent were all critically important in the competition for
record sales. The role of recording company as cultural intermediary had
arrived.

The first threat – radio

The seeds of the first US crisis for the recording industry were sown at the beginning of the third decade of the 20th century. In 1920 there were two important milestones, one commercial, occurring in the US, the other technological, occurring in the UK. The former was the first commercial radio broadcast, and the latter was the first electrical recording from a remote pickup. With these developments, the evolution of the radio era gained pace. Even though the programme content of broadcasts in the first half of the 1920s was quite basic, radio was free, and direct broadcast had a higher sound quality than discs. Early broadcast licences were granted on the basis that records were not played, but this was generally ignored despite the lobbying from music publishers and musicians' unions. Being initially dismissive, then hostile to the new broadcast medium, the recording companies took some time to recognize the associated opportunities and threats which radio and electronic recording presented. Consequently, revenues from players and recordings fell steadily from their 1921 peak.

Eventually Victor and others collaborated to some extent with Bell Laboratories, the pioneers of electronic recording, and with the Radio Corporation of America (RCA), the government-supported company which was at the forefront of radio. The products of these collaborations were new players of electronically recorded discs which also accommodated radios, the most notable being Victor's Orthophonic Victrola and Brunswick's Panatrope. Expectations of these new machines were high. Victor spent $6 million promoting the launch of the new Victrola in November 1925, and within a week had taken orders for $20 million. Subsequently its shares almost doubled in value relative to their price earlier in the year.

With great prescience, or luck, Victor shareholders sold their company to private investors one year later for $40 million, a price which was based on the assumption that the recording industry was still in robust health. Thirteen months after that, in January 1929, Victor was acquired by RCA to form RCA-Victor. RCA was not committed to the recording business, but had an urgent need for Victor's manufacturing plant and its network of dealers in order to satisfy the rampant demand for RCA radios. In October the same year, the stock market crashed spectacularly. Whilst almost all industries suffered badly from the Great Depression, the blow seemed fatal for the recording business. In 1927, sales of discs had been 104 million. By 1932 they had fallen to a mere six million. Sales of players fell from 987,000 to 40,000 over the same period. The cause of the collapse was a combination of much improved radio broadcasting, a lack of support from RCA for the recording business, and the perceived poor value-for-money of expensive discs and players which needed replacing as technology and formats changed. All these factors contributed to a level of consumer dissatisfaction with the recording industry which almost led to its demise.

Recovery led from Europe

The factors which led to US disenchantment with gramophones were not so noticeable in Europe, where competition from radio was much less significant. Britain had been responsible for much of the innovation in classical music recording since Thomas Beecham's pioneering recordings for HMV, and there remained a deep cultural and commercial commitment to recording. Furthermore, Britain still regarded the gramophone as a highly desirable home ornament and cultural accessory. US-led economic depression was similarly biting in Europe, but in 1931 the two major British competitors merged. Columbia and The Gramophone Company (owner of the HMV trademark) combined under a new company, Electric and Musical Industries (EMI), which provided essential scale economies in difficult times. One of EMI's thrifty initiatives was a project which encouraged private subscribers, via societies, to finance a wider repertoire of new classical recordings than would have been otherwise available. EMI's leadership in recording technology was further underpinned when it opened new studios at Abbey Road in 1931, marked by a recording of Edward Elgar conducting the London Symphony Orchestra.

The flow of new and higher quality recordings from Europe kept US interest in the recording business alive. The concept, or rather the marketing, of 'high fidelity'[3] sound also took hold at this time. This was 'more gimmick than gain' (Gelatt 1956, p. 207), given that for most people the new and superior sound source was severely constrained by the low-quality amplification and speakers which were generally available at that time. Nevertheless, the aspiration for 'high fidelity' sold records in great quantities.

Another industrial stimulus imported to the US from the UK was the much lower pricing of records, led by a newcomer, the Decca Company, which in 1934 radically reduced the prices of records from big name artists from 75 cents to 35 cents. This coincided with RCA-Victor's introduction of a cheap disc player attachment to the radio-set, the Duo Jr. which at $16.50 was a small fraction of the price of any previous player. By the mid-1930s, the consumer perception of value-for-money was restored. The US recording industry returned to growth, and domestic record sales were now competing with the sales of records for use in jukeboxes. Jukeboxes had started to become extremely popular in bars and restaurants following the repeal of alcohol prohibition, which had been enforced between 1920 and 1933. Popular music, especially big band jazz music for dancing, played a leading role in the ensuing cultural liberation.

By 1938, RCA-Victor and Decca represented at least 75% of the 33 million records sold in the US that year. Although 33 million was well above the nadir of six million in 1932, it was still well below the late 1920s peak of more than 100 million. Columbia was struggling because it had not benefited from the commercial initiatives of its rivals, nor had it invested in recordings of new repertoire. In an ironic reversal of fortunes, the ailing Columbia Phonograph was bought at the end of 1938 by Columbia Broadcasting System (CBS), the radio network to which Columbia Phonograph had given its name as a short-term rescuer-investor in the fledgling broadcasting company a decade earlier. CBS was now growing quickly, and challenging its main competitor, NBC, which was the broadcasting arm of RCA. Between RCA and CBS, radio broadcasters now controlled the two best-known record companies in the US. Columbia Phonograph revived under its new broadcaster-owners, who recognized that new repertoire was critical to the satisfaction of consumer demand for new recordings in both popular and classical music.

The second threat – war on two fronts

Intense competition between all three companies led to a growth spurt, and in 1941 the US industry finally surpassed its 1920s peak, achieving sales of 127 million discs. The boom was short-lived. America entered the Second World War, and in 1942 the importation of shellac, the primary component of discs, was cut by 70%. Production of electrical goods for civilian consumption was also halted.

As it turned out, war overseas was not the greatest threat for the industry. Four months after the shellac restriction was imposed, a complete cessation of recording activity was imposed from a different source. His name was James Caesar Petrillo, the President of the American Federation of Musicians. The ban covered everyone from jobbing musicians to band leaders and classical conductors. Petrillo had been waging war on recorded music played on the radio and in jukeboxes for more than a decade, and he promoted Sousa's phrase 'canned music' (1906) to great derogatory effect. Petrillo's arguments recalled the days when thousands of restaurants and dance halls had employed their own bands, and radio stations had their own orchestras. Jukeboxes and records played on the radio were therefore bad for the livelihood of musicians, or so it seemed to Petrillo. Petrillo had previously achieved partial successes for his union members, such as the insistence on having a union musician employed in every radio station simply to operate the turntables. But the 1942 strike preventing all recording was more dramatic than anything he had previously achieved. Decca suffered the worst. Being the newest company, they had the smallest catalogue of recordings to fall back on during the ban. Decca was forced to the negotiating table after 13 months, conceding a royalty payable to the AFM on all records. Victor and Columbia gave in after two years, agreeing similar royalties on Armistice Day 1944.

The second recovery

The combination of the recording ban and the shellac restriction meant that huge demand had built up over the war years, which had dramatic consequences when both issues were resolved. In 1946 US annual sales leapt to 275 million discs, and then to 400 million the following year. Sales of disc players in 1947 were 3.4 million. Europe was also innovating in recording technology. One example is full frequency range recording (FFRR), which was developed by English Decca as a by-product of military-commissioned research. But the end of the war led to a more startling revelation for English and American sound engineers, which was the German progress in the development of magnetic tape. Tape had several advantages over disc, not least that it enabled considerably longer continuous recording times. For all the improvements in sound quality, discs had revolved at 78 rpm since the early 1920s and needed to be changed every four minutes, a drawback which was scarcely tolerable for the enjoyment of classical recordings. Tape, by contrast, could run for 30 minutes. Tape had first been developed by a Danish engineer in 1899, but had not been commercialized for musical recordings due to various technical obstacles relating to amplification. These had been solved in the intervening years, and the German machines of the 1940s were technically impressive, but they were also expensive and the tapes were bulky. The commercial potential of tape was identified by the Minnesota Mining and Manufacturing Company (later re-named 3M). 3M was famous for its Scotch Tape brand of adhesive materials and was an innovator in the commercialization of a more practical lightweight magnetic recording tape. By the time Bing Crosby insisted on a move to tape for the recording of his radio shows in 1947, a landslide had started in favour of the tape format for radio productions and master recording.

Tape may well have replaced discs as the preferred consumer format for musical recordings had not Peter Goldmark, a Hungarian-born engineer working in Columbia's laboratory, managed to produce a new long playing disc, the LP Microgroove Record. Columbia recognized the need to maintain innovation leadership, and in 1944 had set Goldmark the challenge of developing a long playing record. By 1947 he produced a disc which maintained the 12-inch standard diameter, but was made of vinylite and played at a slower speed of 33 and 1/3 rpm, and had narrower grooves. This combination meant that it could play for more than 20 minutes each side, with no loss of sound quality. Hoping for industry-wide adoption of the new format, in 1948 Columbia invited its main competitor RCA-Victor to share its system of LP recording. RCA-Victor ignored the invitation and within a year launched its own competing format, the 7-inch, 45 rpm, microgroove record. The advantages of the 45 were its compactness, and the fact that new smaller machines had much faster automatic disc changers. But the playing time was still only four minutes. RCA-Victor was therefore banking on domestic storage convenience being more attractive to consumers than the appeal of longer continuous playing times. The latter was more appealing to the classical market, the former to the mass market for popular songs.

The format battle between the LP and the 45 confused the consumer, and in 1949 sales fell by more than a fifth. It took until 1953 for the format question to be standardized internationally, with the predominance of players which were compatible with both disc sizes and speeds. The formats ultimately became known as 45s or singles, and LPs or albums. With hindsight, it seems inevitable that the industry would eventually settle into maintaining two formats which allowed them to differentiate demand for music, and to market varied repertoire in different ways and at different prices. But that positive outcome was not seen at the time and is a good example of a hindsight rationalization of the intransigence of RCA-Victor and Columbia in refusing to concede format victory to each other. As it turned out, more by necessity than design, the establishment of two ultimately complementary formats meant that the competing corporate technology interests contributed to the shaping of culture, and artists adapted to the available formats. Classical music, Broadway shows, jazz and themed popular music thrived on the LP format. The 45 was better suited to musicians, composers and performers who were more inclined to produce hit songs, or singles. It was only in the late 1960s and 1970s that contemporary artists began to explore the musical possibilities of the LP with longer and more holistic musical creations.

There were further improvements in sound reproduction technology, such as stereophonic sound, which were instrumental in ruthlessly exploiting the aspirational audiophile through sales of ever more expensive players, amplifiers and speakers. However, by the mid-1950s the dominant physical formats for general consumption were now largely settled for the next 20 years.

Symbiosis and prosperity

The birth of the new formats coincided with the so-called 'baby-boom', and in many ways the evolution of singles and albums became culturally and economically intertwined with their human siblings. There was unprecedented economic prosperity in the US and the emergence of 'white' rock and roll from its R&B roots illustrated the cultural exuberance of the time. Radio blossomed as a cultural medium for the young through the rise of the disc jockey, or DJ, as tastemaker and trend-setter.

Competition amongst record companies for new musical talent was fierce, but segmentation of the market meant that new divisions and 'labels' sprang up to accommodate the demand, creating new stars and making healthy profits. Rather than being a threat to music sales, radio had settled into a symbiotic, if sometimes corrupt,[4] relationship with the recording industry. Radio, and the excitement of the top 40 charts, fuelled the desire for music. Tiny transistor radios and small portable 45 players were easily affordable by the early 1960s, and made music an essential and ubiquitous part of the young liberated and expressive lifestyle.

The music industry boomed through the 1960s and 1970s, not only in the US, but also internationally, helped by the rich and diverse cultural contribution from British artists. Growth can be attributed to a number of factors which were technological as well as cultural. One factor was the growing popularity of the LP album as a medium for new musical innovation and artistic expression. Pink Floyd's *Dark Side of the Moon* is an iconic example. Other sources of 1970s success include an explosion of new genres such as reggae, punk, hip hop, and disco on top of the diversifying repertoire within the genres of pop, rock, soul, funk and traditional R&B. International marketing campaigns became powerfully coordinated, leading to the global dominance of Anglo-American repertoire. There were also some hugely successful movie soundtracks, with *Saturday Night Fever* and *Grease* each selling in excess of 25 million units.

As a result of all these factors, the industry achieved a historical peak towards the end of the decade,[5] more or less 100 years after Edison's invention.

1980–1999 – portability and global domination

Though it was before my time in the business, anecdotes[6] from my older colleagues made it clear that in the early 1980s the business suffered something of a post-1970s hangover, marked by falling sales in the early part of the new decade. This has been interpreted as resulting from poor corporate governance in controlling the extravagance and wild excesses of the late 1970s. Other arguments offered for industry decline include a complacent and short-termist approach to the quantity and quality of investment in new artists, and a cyclical cultural lull in the emergence of new talent and social trends.

One further possible cause of short-term decline, though often disputed,[7] was the growing popularity of the musicassette (MC) format. Blank recordable cassettes facilitated home-taping and sharing, thereby arguably 'cannibalizing' the sales of LPs. The UK industry's public education campaign in the 1980s that 'home-taping is killing music' was subsequently ridiculed when the extent of the longer-term contribution of the cassette became more apparent. By the end of the 1970s the radio-cassette player had already become the default standard for in-car audio. Even more significantly, the popularity of the cassette led to the development of the Sony Walkman, one of the most successful products ever produced by the Japanese electronics giant. The Walkman was launched in 1979 and it transformed the consumption of music globally. It made music very portable and very personal, and quickly became a highly desirable lifestyle accessory for hundreds of millions[8] of people. It was the iPod of its day and undoubtedly stimulated the global demand for recorded music, even after taking account of increases in home-taping.

By the time the Walkman was just getting into its stride, another product innovation emerged, driven by technology. In 1982 Sony, in collaboration with Philips, the Dutch electronics giant that had invented the musicassette in the 1960s, launched the compact disc (CD). With a futuristic look, and attributes such as greater robustness, longevity, convenience, and arguably higher sound quality,[9] the CD was successfully marketed as a superior product to both the MC and the LP. Consequently, many library-building consumers re-purchased albums on CD which they already owned on other formats. By the second half of the 1980s, the recorded music business was back on a healthy growth curve which continued right up to the end of the century. This growth allowed the major record companies to invest in efficient large-scale manufacturing plants and global marketing and distribution networks, further strengthening their control of the lion's share of the music value chain. Technology-driven product innovation had seemingly secured the future of the industry, just as it had done repeatedly since the birth of recording technology 100 years earlier.

International corporate empire-building

During the 1980s and 1990s, corporate ownership of the music industry became much more international. In 1986 the German media company Bertelsmann acquired RCA Records (including the Victor legacy) and formed the Bertelsmann Music Group (BMG). In 1987 Sony of Japan acquired CBS Records (including the Columbia US legacy), re-naming it Sony Music Entertainment (SME) in 1991. The Dutch company PolyGram had been especially acquisitive of established independent record labels. I joined the company in early 1992 and the atmosphere was exuberant. I was involved in the acquisitions of Motown and Def Jam, and in the post-acquisition integration to the Group of A&M and Island Records, which had each been acquired in 1989. PolyGram already had an impressive stable of labels which included Phonogram, Mercury, Polydor, Decca, Philips, Verve, London Records and Deutsche Grammophon, the latter being the German company founded by Emile Berliner in 1898. By the late 1990s, PolyGram could claim to be the biggest record company in the world.

Then in late 1998, Seagram CEO Edgar Bronfman Jr. controversially extended the Canadian drinks company's new direction into entertainment when he bought PolyGram from its parent company, Philips, for US$10 billion. PolyGram was merged with Seagram's own recently acquired and re-branded Universal Music Group, which included the labels MCA, Geffen and Interscope Records. The integration of the two companies was successful, and the newly combined powerhouse then represented well over one-quarter of the recorded music market in most territories. At the time of writing, Universal Music Group still retains its market leadership, further consolidating its dominant position with the 2012 acquisition of EMI.[10]

In the late 1990s there was growing pressure in the corporate finance world to maximize the 'new media' value of market-leading entertainment content through consolidation with traditional media and communications companies. This pressure led Bronfman to merge Seagram (including the newly merged Universal Music Group) with the French conglomerate Vivendi in 2000, forming Vivendi-Universal in a transaction valued at US$34 billion.

This sounds like a promising story, at least for corporate shareholder value.[11] However, in June 1999, only six months after the PolyGram acquisition, the peer-to-peer file-sharing service Napster was launched. Although it was widely expected that there would be trouble ahead, no one was gloomy enough to predict just then that 1999 would mark a historical peak of the global recorded music market, nor that a dozen years later the market would have lost more than half of its value. In fact on an inflation-adjusted, per-capita basis, the 2012 US recorded music market was even below half of the value of its previous peak in the late 1970s. Nor was the trauma which lay ahead foreseen one year later when Vivendi-Universal was formed in June 2000. This was only six months after another giant bet on the synergies between new media and old, the record-breaking US$164 billion merger of AOL and Time Warner, which no doubt influenced Vivendi. Vivendi was the new corporate brand name for a very old French water and utilities institution, which had diversified into French media and telecoms in the 1990s. Vivendi had a maverick CEO, Jean-Marie Messier, who was sure that owners of wires and pipes needed to control attractive content in order to maximize their value. Like Voxstar and Vizzavi, and indeed AOL Time Warner, Messier ultimately paid the price of precociousness and huge ambition. Two years after the creation of Vivendi-Universal, Messier was ignobly dismissed amid accusations (and later criminal charges) of misleading investors and destroying shareholder value.

1999–2003 – the file-sharing pandemic

Napster, in its unauthorized form, closed in 2001, only two years after its launch and following lawsuits from the recording industry. However Napster had opened a floodgate, and similar services such as Kazaa, Gnutella, BitTorrent and RapidShare emerged and grew for several years at a rate which outpaced the industry's efforts to have them shut down. Napster and its successors stole the headlines as the nemesis to the new-millennial hubris of those who would tightly control the dissemination of music, but the rapid growth of unauthorized peer-to-peer file-sharing is certainly not the only cause of the collapse of corporate shareholder value in the music business. It is true that Napster had unsettled many senior executives who were understandably anxious and duty-bound to protect the value of corporate assets represented by intellectual property rights, catalogues and large un-recouped advances to artists. After all, these were the products of decades of investment in the development of artists' careers, and they underpinned corporate balance sheets and share prices. Nevertheless, despite this anxiety, there was also within the company a counterbalancing feeling of excitement about new technologies.

This more positive feeling is rarely written about because it does not fit neatly with the dominant blog-narrative of *schadenfreude*; i.e., that the record industry somehow got its just desert for being managed by greedy, complacent, white, middle-aged male Luddites working within a system which often appears to waste more talent than it develops. Such a resentful view is understandable, but does not accommodate my own counterbalancing observations that music companies are mostly populated with people who choose to work in the industry because they love to be involved with the development of artists, and with the processes of discovery, production and dissemination of great music. Whilst some of those people are characterized by arrogance, egotism and vanity, many more work with empathy towards artists, and with humility and dedication. Most are bound by an interest in playing a role, directly or indirectly, in the success of talented and/or glamorous people, and in bringing musical delight to a big audience. Though we felt anxious about the disruption of new technologies, record company employees equally felt excitement and anticipation that we were on the front line of profound change which went beyond the prevailing value-chain of music production and distribution. In the early years at least, there was an assumption that the record company had a role, and even a responsibility, to be a pioneer of new forms of music discovery and consumption; not by being a barrier to new technology, but by promoting and co-developing it, continuing the 20th-century industrial legacy.

The problem is that the recording industry has always had a dual culture, a split personality which makes it difficult to reconcile the rather different roles of scientist-inventor and cultural patron-curator. But more of that later...

The more immediate and transparent problem was that within months of Napster launching, a frighteningly large proportion of all the music ever recorded, including new recordings before they were officially released, was available on the internet for free. It was available not just to experts and to early adopters of technology, but also available relatively easily around the world to anyone who had internet access; or anyone who had a friend or family member with internet access and a CD burner. The audio CD, which had been the saviour of the industry in the 1980s, was a digital product which could be copied and was now perceived as an insecure product. Prior to Napster, it was reasonable to assume that corporate strategies could be adapted and optimized in ways which could manage the risk of unauthorized copying as the threat evolved. It seemed reasonable to assume that this would be achieved through the control of copy-enabling and disabling software and hardware, and through the enforcement of patents and copyright. With the benefit of hindsight, this seems to be a glaringly obvious corporate misjudgement, but even the most prescient of companies could not have recognized, before the mid-1990s at the earliest, the scale and immediacy of risk which the unprotected CD format[12] presented. There were two important flaws in the imagined scenarios of record companies. The first was a failure to notice that consumers had been copying CDs to their computer hard-drives since the mid-1990s, often just for personal use, and that this attribute of the CD was one of its appealing qualities. The second flaw was the deep-seated presumption that very high fidelity audio reproduction would always be a prerequisite for music consumers.

'What a fuss people make about fidelity' (Wilde 1891)

The pursuit of natural 'lifelike' sound reproduction was the original mission of recording companies, and gave them their right to participate in the revenue potential of musicians and composers. It brought the great classical performers such as Caruso, Melba, Beecham and Elgar into the living room. In his book *Capturing Sound*, Mark Katz points out that for more than a century 'a discourse of realism has reinforced the idea of recorded sound as a mirror of sonic reality, while at the same time obscuring the true impact of the technology' (Katz 2004, p. 1). By 'true impact' he means that recorded sounds, as *mediated* sounds in their various transformations, have their own independent properties and characteristics which encourage new creative practices and aesthetics. The new practices are in musical performance, in production, in the creation of new genres, in listening behaviours, and in the socialization of music. The idea of 'natural' or 'lifelike' sound now has little meaning, as the vast majority of what we hear has been mediated by technology,

even in 'live' performance which is so frequently assisted by recording technology in one form or another. The recording industry has
long understood how to take advantage of some of these technological
impacts, but it severely underestimated the degree to which consumers
were happy to sacrifice a relatively small amount of reproductive purity
for the benefits of choice, convenience and discovery.

Enter MP3

The catalyst of the highly disruptive file-sharing phenomenon was the
audio compression format known as MP3, a short form of MPEG-1, or
MPEG-2, Audio Layer 3. MP3 is a product of a long history of scientific
research in telephony and psychoacoustics begun by AT&T's Bell Laboratories in the 1920s. Of most relevance here is the concept of perceptual
coding. This is the name given to the solution which makes it possible to
exclude, for the purpose of efficient transmission, those parts of sound
which are not likely to be audible to the human ear. The industrial goal
for telecommunications companies is to maximize the capacity of their
systems and infrastructure by *minimizing* the amount of data required to
transmit a signal of acceptable quality. It is interesting to reflect on how
this is a quite different industrial goal from the music industry which
prioritized audio reproductive quality over data storage and transmission efficiency. This subtle divergence of technical priorities gives some
clue as to why MP3 was not perceived as a threat to the music industry
until it was too late.

From the 1890s to the 1980s, the pursuit of high fidelity (hi-fi) in
sound reproduction was the common goal of engineers, audiophiles and
corporate marketeers alike. In the early decades of the 20th century, the
profits in the recording value chain were driven more by sales of players
than by records. Home stereo sound systems with ever higher specifications and promises of higher fidelity generated billions of dollars
for the electronics industry, and the perception of high sound quality was a key element in the success of the audio CD format. In the
1990s, minimizing the file sizes which created such high quality was
not the biggest industrial priority. It might be said that the electronics and recording industries misjudged the tolerances of the 'public ear'
as it relates to music consumption. For many years they successfully
marketed products which were technically over-specified for much of
the public's listening behaviour. It only became apparent following the
success of the iPod and iTunes, but the public were prepared to sacrifice high-end reproductive quality in return for the greater convenience,
accessibility and personalized experiences made possible by small digital

files. Given the cleverness of perceptual coding, combined with the high proportion of music which is consumed via cheap button headphones or in noisy environments, most consumers are not even aware of the sacrifice. But this level of public tolerance for audio compression was not well understood in the 1990s. Audiophiles regarded high levels of compression as technologically regressive; artists and A&R executives regarded it as aesthetically degrading.[13] A good deal of pride would have to be swallowed before the industry could accept that high fidelity was not as important as they thought.

The end of the industrial symbiosis between music and technology

In order to understand the evolution of the MP3 format and the way it blindsided the music industry, it is particularly relevant to understand the music technology strategies which were being pursued by Philips and Sony in the late 1980s and early 1990s. As successful as the CD was, its size was regarded as being too large to support a really convenient mobile player which could be attached to the human body whilst physically active. The cassette Walkman was a robust portable device but it was looking very old-fashioned in a digital world. Sony and Philips, who had collaborated so lucratively on the establishment of the CD audio standard, were now in a highly competitive race, working separately on digital portable solutions to replace the cassette and its related hardware products. Sony's solution was the MiniDisc, which resembled a smaller version of the CD. Philips solution was the Digital Compact Cassette (DCC) which resembled a smarter hi-tech version of the analogue cassette. Both launched in 1992, and both ultimately failed to make a serious commercial impact. The DCC had disappeared by 1997. The MiniDisc was more popular, especially in Sony's home market of Japan. However, it never captured the success of either the CD format or the cassette Walkman and was finally discontinued in 2013. Unlike Philips, Sony did experiment with a portable digital media player, and with direct-to-consumer online music sales through its Sony Connect programme, launched in 2004. Connect failed, and Sony abandoned it four years later, recognizing that its consumer proposition was too restrictive and simply not as compelling as the alternatives, especially iTunes.

Although they considered themselves to be pursuing *digital* strategies, Sony and Philips had been continuing to assume they could control the consumption experience by concentrating primarily on *physical* products: discs, tapes and associated hardware. Their content-owning

subsidiaries referred to themselves as *record* companies, being insepara-
bly identified with their physical products. In a way, the whole industry
was cognitively constrained, carrying almost 100 years of 'common
sense' that high fidelity and well-packaged physical products would
always be critical attributes in consumers' perception of value, i.e. what
they would be prepared to pay for. Though Philips and Sony were fully
aware of the evolution of much smaller compressed digital files such
as MP3, they did not perceive them as being a direct threat, as they
seemed to be insubstantial products which were clearly of insufficient
audio quality to compete with the CD or the MiniDisc. It is easy to
look back on those years and wonder at the lack of vision which might
have seen that digital content was on the verge of breaking free from
its plastic containers. But a century of common sense takes a while to
overturn. In his excellent book, *MP3: The Meaning of a Format,* Jonathan
Sterne comments:

> The relative absence of innovation in the mainstream recording
> industry is crucial to the MP3 story. [...] In its formative years online
> music was not the province of the recording industry, which had
> hitherto done a fairly good job of controlling its distribution chan-
> nels. Online music – which was at its core a mode of distribution,
> a relation to infrastructure – was instead the province of companies
> like Fraunhofer and Philips, Microsoft and RealNetworks.
>
> (Sterne 2012, pp. 203–204)

It is interesting that Sterne does not equate the 'province of Philips'
with the province of the recording industry, despite Philips being so
instrumental in the growth of the recording industry through its development of the cassette and the CD, and through its majority ownership
of market-leading music label group, PolyGram. Of even more interest, however, is to appreciate the irony that Philips was so actively
involved in MPEG *without* seeing the scope of its impact on the music
business.

The role of MPEG

MPEG stands for the Motion Picture Experts Group, a collaborative
international network of scientist–technicians who shared an aspirational goal of establishing global technical standards. MPEG was established in 1988 as an offshoot of JPEG, which set standards for still
photographic images. The network is truly international,[14] and was

motivated by the desire to avoid the huge industrial inefficiency and consumer frustrations caused by competing standards in audio-visual and communications systems and infrastructure. It also aimed to reduce the risk of second-rate technical solutions becoming dominant. At the time of MPEG's formation, full bandwidth digital CD audio could not be combined with digital video onto a disc, nor was it suitable for phone lines or radio. These constraints hindered both the technical progress, and the commercial potential, of enhanced audio-visual quality in home video, broadcasting and telecommunications. MPEG therefore set about establishing digital audio compression standards.

Despite MPEG's aspirations to suppress corporate competitive agendas, two competing coding-decoding ('codec') standards emerged early on. In an effort to mediate conflicting corporate interests, both codecs were incorporated as options in the standard, and named as MPEG Layer 2 and MPEG Layer 3. Layer 2 was less complex but also less efficient. It was backed by Philips and Panasonic. Layer 3 was backed by the research engineers at AT&T, the German Fraunhofer Institute and others.

Being one of the pioneers of the CD and of the first optical videodisc formats in the 1970s, Philips was a highly credible expert member of MPEG. In the still undeveloped digital home video market, Philips was keen to replicate its CD patent success in the global audio market by leading the process in developing the next generation of video discs. As a company with a powerful presence in radio and TV products, Philips also had a strong strategic interest in pioneering the standards in satellite television and digital audio broadcasting. Audio compression was a key element of these standards. By 1995, Philips had seemingly got its way, with MPEG Layer 2 being adopted for digital audio and TV broadcasting, and for digital video discs (DVDs).

Before commenting further on Philips' apparent success and the subsequent strategies of Philips and Sony, it is important to understand what happened next to the 'losing' format, Layer 3. The Fraunhofer Institute, still convinced of the superiority of Layer 3, continued its marketing of the alternative standard in what seemed at the time to be niche channels. It became the preferred standard for the Internet Underground Music Archive (IUMA) which is acknowledged to be the first major player in online music distribution, being set up in 1993. As recounted by Sterne (2012), IUMA switched from Layer 2 to Layer 3 not only because of its higher quality with smaller file sizes, but because it became freely available. This availability was unauthorized, being distributed by an Australian hacker who in 1995 acquired Fraunhofer's program with a stolen credit card, and adapted the interface, redistributing it for free under the name 'thank-you Fraunhofer'. It was around this time that Fraunhofer re-branded MPEG Layer 3 with the name MP3.

In addition to being the format of choice for technically savvy musicians, active music consumers and hackers, Microsoft adopted MP3 in 1995 within applications bundled with its operating system. By 1999, the MP3 format was also compatible with Apple's Quicktime. A further boost came from the launch of various personal, portable, digital audio players from 1997 onwards. The most well known was Diamond Multimedia's Rio, being the first of its kind to attract the attention of music industry lawyers, and Diamond successfully defended itself against the 1998 lawsuit brought by the Recording Industry Association of America (RIAA).

A perfect storm

Thus, between 1995 and 1999, a series of technology developments conspired to create the perfect storm for the mainstream recording industry: the hacking of the MP3 encoding program and its unauthorized free distribution; the broad acceptance of the MP3 sound compression standard for file transfers amongst communities of technicians, musicians and highly active music consumers; the incorporation of MP3-compatible applications in both Windows and Apple computers; the rapid domestic market penetration of personal computers with CD drives which allowed the widespread 'ripping' and 'burning' of CDs; the appearance of portable digital media players; and the development and rapid viral spread of peer-to-peer file-sharing technology. Finally, and perhaps most damaging, was that the record labels had corporate parents who were distracted by other priorities. During these years, Sony and Philips were engaged in winning other media format battles which they perceived to be more lucrative and which went well beyond the domain of recorded music.

From the mid-1990s onwards, there was in both Philips and Sony a growing strategic misalignment between their electronics divisions and their music divisions. Sony had a number of conflicts where its music division was in litigation with its other product divisions. The latter endeavoured, rationally enough, to exploit new consumer demand from the emerging technologies, such as internet radio services and CD-burning hardware, unconcerned as to the detrimental effect it might have on Sony Music. Similar tensions emerged between

PolyGram and its corporate parent Philips. I joined PolyGram in 1992, the same year in which Philips launched the ill-fated DCC. Whilst Philips and PolyGram had distinct cultures, they were both Dutch and there was still a strong mutual respect and confidence in the symbiotic relationship. Jan Timmer, Philips' CEO from 1990 to 1996, had been in charge of PolyGram from the difficult early 1980s, and he is credited with rescuing not only PolyGram, but also the whole music business through his role in the successful launch of the CD. There was, therefore, loyalty and strategic understanding between Philips and PolyGram during Timmer's tenure as Philips CEO. However, by then Philips was suffering financially. It had grown to become a complex and unwieldy global company of over 250,000 employees, and a reputation for great technology innovation rather than for smart marketing strategies. In 1996, Cor Boonstra took over from Jan Timmer, and was the first Philips CEO not to have developed his career within the Philips group. He came from American food group Sara Lee and had no experience or empathy with the entertainment business. He began rationalizing the many divisions and companies within the Group, and successfully transformed Philips into a company with more of a consumer-electronics brand focus under the campaign 'Let's make things better'.

As regards music strategy, the failure of the DCC format was only one of several factors which had a detrimental impact on the relationship between Philips and PolyGram under Boonstra's tenure. Other factors were Boonstra's nervousness in not knowing the entertainment business, and having to rely on 'mercurial and difficult management'.[15] He was also anxious about PolyGram ploughing its substantial cash-flow into its fledgling film business, and he was nervous about the threat to the music business posed by the internet. In his account of the demise of PolyGram's film business following its acquisition by Seagram, Michael Kuhn gives some insights into this important phase in the history of the relationship between music content and technology. Recounting a late 1997 meeting between PolyGram executives and Boonstra, he writes:

> We decided to have it out. We convened a meeting in a country house hotel, south of London. [. . .] After a pleasant dinner we set out to discover the truth. We said that we thought that he [Boonstra] was not particularly interested in PolyGram as such. He said that was correct. We then discussed whether he would like Philips to get out of their investment if they could. He said he would. We then devised a plan

whereby we would each explore various exit scenarios with Philips and reconvene in due course.

(Kuhn 2003, p. 92)

Unknown to PolyGram's management, Philips already had a process underway and was in talks with Seagram. The sale was announced in May 1998, only a few months after that meeting.

I draw attention to this event because it is relevant to describe the emotions within PolyGram when it was announced that we were to be sold. I had attended a PolyGram senior international finance meeting a few days before the announcement. At that meeting, we expressed great concern that the Group (Philips and PolyGram) did not appear to have a strategy regarding new digital technology and the internet. The DCC had been quietly withdrawn from the market 18 months earlier, and since then there was a sense that Philips, under new direction, was giving up on its long relationship with recorded music technology and products.[16] Despite this premonition, there was a profound sense of shock, betrayal and sadness when the news of the disposal finally broke. PolyGram was a company which discovered, developed and managed talent, and which promoted, manufactured and distributed physical products. But as successful as PolyGram was at those things, it did not have the expertise, structure, resources and culture of technical innovation with which to be a major player in the domain of the new type of solutions which were to shape the future consumption of music. More worryingly, neither did our new owners, Seagram.

It is tempting to imagine that if Philips had, through its active role in MPEG, better understood the potential life of digital music files beyond their physical containers, things might have turned out differently. In parallel with MPEG, Sony and Philips were each pursuing independent research and development of media compression as part of their MiniDisc and DCC strategies. Each had a proprietary codec which, if it had come to market dominance instead of MP3, might have given either company, and the industry in general, more control over internet file transfers, relegating alternative unauthorized formats to the fringes. As a former PolyGram manager, I therefore feel that a huge industrial opportunity was missed. By contrast, as an active music consumer, I feel that the MP3 story is a serendipitous one: it is difficult to imagine that, had it retained tight control of digital music file distribution, the recording industry would have created a music consumption experience anywhere close to the breadth and flexibility of what is available today.

2003 to? – the golden age of Apple

The withdrawal by Sony and Philips from the device market eliminated two major competitive threats to new entrants, and thereby created an enormous opportunity. In his book *Good Strategy/Bad Strategy*, Richard Rumelt (2011) recalls a conversation he had with Steve Jobs in 1998, not long after Jobs had returned to Apple to save it from near bankruptcy. He asked Jobs what his strategy was for the longer term, following his initial cuts and refocusing of Apple as a niche player in the global computer market. Jobs replied that he was just 'going to wait for the next big thing' (p. 14). This comment was made the same year in which portable digital music players arrived on the market in notable quantities from a variety of specialist manufacturers such as Audible, Diamond and Creative. Jobs saw the opportunity but he did not rush to market. He recognized that the experience of digitizing and organizing one's music collection was a labour-intensive exercise. It also required technical knowledge and therefore, for a while at least, it would be limited to the domain of early technology-adopting, and highly active music consumers. Any solution which might appeal to a more general consumer would have to make the whole experience much easier and more enjoyable. It required thought, care and time to craft the optimal design.

Apple launched the first iPod three years later in 2001, but it was only when a Microsoft Windows-compatible version of iTunes was released in October 2003 that the mass market of music consumers took notice. It had therefore taken ten years from the first online distribution of music by IUMA, five years from the arrival of the first portable digital music players, and four years from Napster triggering the file-sharing revolution, before a compelling, legal and authorized digital online music offering finally crystallized. Apple delivered what millions of consumers craved, even if most of them didn't know it yet: a holistic and pleasurable solution for organizing, personalizing, discovering and rapidly accessing almost all the music most people would ever want, all wrapped up in an elegantly designed, intuitive-to-use portable device which synchronized simply and reliably with one's computer. As of September 2012,[17] Apple had sold 350 million units of the various iPod models. For the purpose of illustration of the comparative scale of revenues, this broadly equates to an average annual revenue of $5 billion. When added to another $5 billion of music revenues from iTunes, this means that Apple has typically generated from music[18] around twice the annual revenues of market-leading music company Universal Music Group.

As early as 2006, Apple had already grown from having no previous music retailing presence to become the single biggest music retailer in the world, this market share being amplified by the fact that sales of CDs were collapsing in most markets. Apple's dominance of music retailing was further secured with the launches of the iPhone (2006) and iPad (2010) which embedded the Apple media platform even more deeply and widely. Record companies had played no part in the development of this extraordinary new business model for music consumption, other than to license their recordings to iTunes. Even this minimal level of collaboration was conceded with extreme reluctance in some cases. Many in the recording industry, including some major artists who withheld their permission, thought that the brutally simple pricing of 99 cents for all songs, combined with the unbundling of the album format which allowed consumers to cherry-pick individual songs, had the effect of commoditizing music. It therefore constituted a disastrous strategy which diluted economic value. Theoretically they had a strong argument, but Jobs had a stronger one to which he held firm: that simplicity, flexibility and choice were the key attributes which secured the mass-market adoption of iTunes. Competing economic pricing theories of value-maximization had no relevance here because the recording industry had no credible alternative business models. They therefore had no choice but to accept Apple's terms of trade. Though there was something of a love–hate relationship between Apple and the record companies for many years, only die-hard music industry veterans would now fail to acknowledge that Apple prevented an even sharper decline in the recorded music industry in the first decade of the new millennium.

Subscription services

In early 2014 the ten-year-long boom of purchased digital downloads, led and dominated by iTunes, showed signs of waning.[19] Generation X and Baby Boomers had to work hard to discover and access the music they wanted, and when they found it, they wanted to own it. Steve Jobs understood that very well. But to the younger end of Generation Y, or the Millennials, the benefits of owning digital products, when they are so ubiquitous, plentiful, and cumbersome to store and manage oneself, are nowhere near as compelling.

Though many alternative digital music services and models which are not based on ownership have long been established, they still have minor market shares. Internet radio services, which offer free streamed songs tailored to individual consumer preference and recommendation, seem successful measured by advertising revenues and by usage. Leading provider Pandora claims to have 77 million active listeners[20], but remains loss-making. The phone companies in particular failed to collaborate effectively or innovatively with the music industry in order to create models which could seriously compete with Apple and others. The causes of this will be explored in Part III of the book.

The website thecontentmap.com claims that there are more legal online music services in the UK than anywhere else in the world, listing 71 consumer offerings. In revenue terms, the most successful, after iTunes, is the subscription service Spotify, launched in 2008 from Stockholm. Deezer, originating in Paris, offers a similar service and has had similar international success in attracting paid subscribers. Subscription services were amongst the first credible online music offerings (e.g. Rhapsody in 2001), but as of 2012 they had failed to make a dent in iTunes' globally dominant[21] share of digital music spending. More recently, Google demonstrated that it was treating the music subscription market as a growth area with the announcement in 2013 of its All Access service. The service builds on its Google Play Music platform launched two years earlier, which allows subscribers to store up to 20,000 of their own songs in a cloud-hosted locker. Google-owned YouTube has become one of the primary free streaming sites for music discovery and is itself scheduled to launch a subscription service, Music Pass, later in 2014.

Given the success of long-established subscription models in the TV market, it is curious they are not yet as successful for music. They give more choice and access to a vastly bigger catalogue than an individual could reasonably own, and they provide a risk-free browsing, discovery and storage experience. At current pricing levels, they also present a very compelling economic argument for the active music consumer who would otherwise buy more than a dozen albums a year. However, the music industry has long held an antipathy towards the subscription model, feeling that 'renting' music rather than buying it outright could further commoditize[22] the supply of music, diminishing consumer commitment to music purchasing, and reducing fan loyalty to individual artists. Record companies were also anxious that they would lose visibility of the detailed sales and consumption data of their products, which has implications for marketing and promotion, and for artist royalty accountability. Furthermore, on the technical side, subscription services have suffered real difficulties in infrastructure, interoperability and bandwidth reliability. These were early obstacles to creating a portable subscription option which is both easy and reliable for the consumer, and secure for the content-owners. Prudence was the priority in the early years, and the consequentially limited consumer offering was self-defeating. Most of those technical problems have now been overcome, but the one price 'all you can eat' model is still too crude and economically inefficient to exploit a huge and diverse market. Broadly speaking, the £9.99 a month model is a great bargain for the top 15% of most voracious music consumers, but a rather expensive luxury for the rest. Until there is mass-market adoption of subscription services, which will then allow price differentiation amongst different types of consumer, the model will continue to have more promotional value than commercial value, and will be economically dilutive[23] for the artists and for the record industry.

Despite these continuing obstacles, for more than ten years I have been predicting that music consumption via an access model, in one form or another, will become more popular than individual digital products (tracks or albums) to be bought and owned. Steve Jobs was always strongly opposed to the subscription model based on his assumption of the existence of a deeply embedded, developed-world construct of music as something to be collected and owned.[24] For many years, Jobs' prejudice alone was a significant obstacle to the development of subscription models, at least until his death in 2011. Subsequently Apple seems to have seen the dangers of Jobs' legacy prejudice towards subscription services, and recognized the risk that the iTunes download store will increasingly struggle to compete with other models of consumption. Consequently, in 2013 the company launched iTunes Radio which has an offering similar to Pandora, but has thus far threatened neither Pandora's nor Spotify's popularity.

In a bolder and more significant move, which implicitly acknowledged that it might have missed the boat on subscriptions, Apple announced in May 2014 its biggest ever acquisition, Beats Electronics, for $3billion. Beats was formed by two individuals with immense credibility and influence in the music world: pioneering hip-hop producer Andre Young (Dr. Dre), and Interscope Records chairman, Jimmy Iovine. Beats currently has only a fledgling subscription offering, but Apple is betting on Beats helping it become the first subscription service to 'get it right'.[25] It will rely on the potential of Beats' more human-driven curation service, rather than the algorithmic recommendations of its competitors, to convert Apple's 500 million[26] or more worldwide users to new models of consumption within its own controlled platform. In doing so, it indicates a rebalancing of Apple's content strategy towards curation and recommendation.

The other, equally significant, element of the acquisition is Beats' leading position in the US market for premium headphones, where it generates over $1 billion of annual revenue from products whose design and brand is beloved by celebrity artists and sports stars. Premium headphones are categorized as those which cost more than $100, and which generally comprise large on-ear or over-ear solutions, often with wireless and noise-cancelling features. Until relatively recently, such specifications were limited to the exclusive domain of audiophiles and business travellers – large headphones being seen as too cumbersome to be chic or cool, despite their superior sound reproduction. As a result, lower quality 'ear buds' or in-ear button headphones have dominated the mobile listening experience for over 30 years. Beats Electronics was reportedly[27] formed in order to address Dr. Dre's dismay at this cultural tolerance of low-quality sound, and the company's products have been changing aesthetic perceptions of larger headphones for the past five years. High fidelity may at last be fighting back, and reclaiming its historical place in the value-chain of recorded music. This collaboration between Apple and Beats Electronics may therefore transform not only the subscription market, but also, for many people, the whole design and experience of listening to music on portable devices.

However, for the moment at least, Spotify still looks like the most promising music subscription service. It claims[28] to have 40 million users as of May 2014, of whom more than ten million are paying customers, the remainder accepting advertising in exchange for the free service. Despite this success, Spotify still represents less than one-fifth[29] of the music revenues generated by iTunes, and it remains loss-making. Competitive threats from Apple/Beats, Google and YouTube may well transform the subscription landscape quite quickly, as there are huge strategic gains (beyond the mere music revenues) for those who can create the most compelling and loyalty-inducing offering to the consumer. It therefore seems likely that Spotify, Pandora, and Deezer will remain highly desirable acquisition targets, not just for Google or Apple, but for other giants such as Microsoft, Amazon and Facebook who may also place a broader strategic premium on the benefits of being dominant players in the future of music curation, recommendation and delivery services.

Where next for the record companies?

There is a tendency, especially amongst people over the age of 40, to think that the shrinkage of the recorded music market is a sad and retrograde development in cultural history. I used to feel that way myself, but have subsequently come to the conclusion that 20th-century music consumption may ultimately be seen to be the cultural anomaly. Prior to the phonograph, music was an intrinsically social phenomenon, being written and performed for live audiences, whether the purpose was ritual, political, educational or simply for entertainment. Solitary listening is far from being a timeless human practice, but over the course of the 20th century, recording technology converted the listening experience into a more private affair which we now take for granted as a fundamental human pleasure. However, this wonderfully illustrative article in a 1923 edition of *Gramophone* magazine reminds us how unnatural and shocking it was to encounter someone listening to music at home, on their own:

> You would look twice to see whether some other person were not hidden in some corner of the room, and if you found no such one would painfully blush, as if you had discovered your friend sniffing cocaine, emptying a bottle of whisky, or plaiting straws in his hair...I fear that if I were discovered listening to the Fifth Symphony without a chaperon to guarantee my sanity, my friends would fall away with grievous shaking of the head.
>
> (Williams 1923 cited in Katz 2004, p. 17)

One might view new media and technology in the 21st century as stimulating a shift in the cultural role of music back towards its social origins, a shift to which record companies have taken some time to adjust. But they are adjusting.

It is easy to romanticize the grass-roots revolutionary aspects of the MP3 story, or to indulge in a nostalgic lament for the collapse of a long and successful marriage between the inventors who industrialized recording technology and the cultural intermediaries who nurtured and commercialized the careers of artists and tastemakers. The marriage was made at the turn of the previous century with names such as Victor, Columbia, His Master's Voice and Deutsche Grammophon. All these names can be traced by corporate lineage to the two remaining market leaders, Universal and Sony,[30] the businesses which today continue to control well over half of the diminished global recorded music market. They may have ceded their legacy roles as technology hardware innovators to a whole new species of giant corporate media innovators, yet despite the apocalyptic discourse surrounding the music business, it is, in the aggregate, in much better health than is popularly reported, when the broader social functions of music are taken into account.

It is true that on an inflation-adjusted basis, the global *recorded* music market is well under half of its peak in 1999. Recent data[31] suggest that 12 years of decline may be finally at an end. Most companies have adjusted their structures and overheads to accommodate the changing economics, maintaining respectable bottom line profit margin percentages. Furthermore, it is generally much less clearly and less frequently reported that the fall in sales of recordings has been largely offset by growth in other sources of revenue, especially live performance.[32] It is also not well understood that recorded music data are often reported separately from music publishing data, even though music businesses often run the two activities in parallel, albeit with separate management and distinct organizational cultures. Music publishing is the part of the business which deals in the compositional rights as distinct from the recording rights. It is a less capital-intensive business, as its role is more focused on collecting revenues from all sources, rather than investing in specific parts of the many forms of music exploitation. Publishing is the discrete, low profile and lower risk part of the music business, and has been consistently profitable throughout the new millennium value-chain disruption, as its revenue streams are more evenly spread. It will remain lucrative, assuming copyright law remains largely intact and enforceable.

The music market is actually made up of many significant revenue streams. These include licensing of recordings (for advertising, movies, TV and games), live performance, broadcast revenues, merchandising, sponsorship, brand marketing, composition/home recording software, sheet music and instrument sales. The complexity inherent in these various rights and business models creates confusion in the public mind about the health of the broader music industry, especially given the piecemeal reporting of market data, and the lack of understanding about the relationship between the publishing and recording business models.

Record companies traditionally did not participate in live performance, merchandising and sponsorship, viewing them as ancillary markets. In the case of developing artists, live tour support is in fact a common promotional *expense* for record companies, rather than a revenue stream. In recent years record companies have been increasingly challenging these strategies and their own contractual arrangements with artists and concert promoters, so that they can participate in revenues which are arguably directly related to their investments. In this respect, it will only be a matter of time before the increasingly narrow terms 'recording' contract and 'record' companies become obsolete, as music businesses extend broader services to artists. Similarly, consumers will become oblivious to the distinction of separately purchased recorded music products within their overall cross-media, multi-channel consumption of music in all its forms. Music businesses are nevertheless reluctant to give up the term 'recorded music market' as the ability to demonstrate the shrinkage of this market is a key argument in the industry's lobbying for stricter statutory copyright-infringement protection.

There are two critical factors which will determine the economic future of the organizations which we still nostalgically refer to as record companies. They are the same factors which will determine the fate of most cultural industries. The first factor has a cultural dimension, is service-driven, and transcends whichever policies and laws prevail. The second is entirely dependent on law and policy reform. Both factors are deeply and firmly socially constructed, historically passing for common-sense ways of organizing the cultural industries. Their claims to common sense have, however, been fiercely contested in the new millennium. Being linguistically constructed, they are highly dependent upon the power of narrative, rhetoric and metaphor, and thus lend themselves to discursive methods of analysis.

The first factor I will call *expert services*. It assumes the retention by music business people of their expertise and extensive resources in the discovery, nurturing, development and bespoke promotion of exceptional talent. This includes their ability to detect, shape and exploit cultural trends through a variety of media, products and services. They will need to continue to master new skills, especially regarding new ways to exploit, directly or through collaboration, the social role of music. These embrace what Kevin Kelly (2008) in his article 'Better than Free' calls his eight inimitable 'generatives' of consumer value, namely: immediacy, personalization, interpretation, authenticity, accessibility, findability, embodiment and patronage. No matter how easy it may have become for musicians to record and to present their music online, nor how efficient and technically sophisticated the curation and recommendation algorithms and other grapevine tools are, the skills of the cultural intermediary will thrive because they are ancient, subtle and complex crafts which underpin mutually productive social roles and identities. If anything, they will be more valued than ever in a world of proliferating and competing products. A growing taste for more local and organically grown talent untainted by big corporate exploitation favours the small independent entrepreneur. At the other end of the spectrum, there is seemingly no diminishment in the public appetite for easily discoverable and digestible hits, style icons and branded glamour, which the bigger corporations are so expert at providing. In both cases, artists are generally disinclined to spend their time being marketeers, entrepreneurs, administrators and sophisticated self-promoters, even now when many of the previous obstacles to them playing these roles have been removed. It therefore seems reasonable to assume that music businesses (rather than record companies), along with other cultural industries, will continue to adapt and master the new media tools and technologies which will help them retain their positions as patrons and curators, even if they neither own nor control the technology for production and dissemination.

Whilst the provision of *expert services* will determine whether the businesses formerly known as record companies continue to have any role in the future value chain of music, a second critical factor will influence the scale and scope of their activities.

That second critical factor, which I will refer to as *capital protection*, is the extent to which intellectual property law, and copyright in particular, might be reformed either in favour or against the current content-owning oligopolies. This is similarly not just an issue for the music industry, but for all the traditional cultural industries, such as book publishing and the film industry. Reform has already proved itself to be a painfully slow political process, not helped by the fact that, in the UK alone, there have been six successive ministers for intellectual property since I interviewed the first one in 2007. There have also been several government-commissioned reviews so far in the new millennium, all more or less commonly concluding that the current law is less than optimal with regard to fostering innovation. Complexity, ignorance and fear all conspire to create obstacles for the development of new products and services which rely on intellectual property. Angry demands to liberate culture from its old captors are inextricably bound up with fears that the 'wrong' kind of liberalization will favour new media giants whose 'big data' armouries may already be creating new monopolies through their control and non-remunerating exploitation of online identities, behaviours and images.

New legislation is sober, modest and unambitious in its reforming scope, but even so has still not made any significant impact on the power relations between content providers and other economic participants, largely due to the rhetorical minefield which it is endeavouring to cross, and to the political preference for industry-brokered voluntary solutions, rather than imposed statutory ones. France and

Germany are more legislatively conservative and protective of their legacy cultures than the UK, and France was the first to implement (and to withdraw) controversial anti-infringement measures which can lead to a person's internet connection being cut off. The continental Europeans are also particularly fearful of the cultural and economic impact of Google, Facebook, Apple and Amazon. The US government is relatively less concerned about cultural protection measures, but is more vulnerable to being swayed by powerful corporate interests, both new and old. The traditional content industries won hard-lobbied pre-Napster victories with the Copyright Term Extension Act (1998) and the protective anti-circumvention measures contained within the Digital Millennium Copyright Act (1998), but since then there has been more deadlock than progress in legislative changes, along with a tsunami of litigation by rights-holders. There is broad acknowledgement of a strong case for copyright reform, with the US Register of Copyrights calling for 'the next great copyright act' recognizing that 'if one needs an army of lawyers to understand the basic precepts of the law, then it's time for a new law' (Pallante 2013).

The following chapters examine these critical factors of expert services and capital protection, with the aim of contributing new insights into the obstacles and opportunities for evolution and reform in the music industry, and the cultural industries as a whole.

Part II
Stakeholder Voices

Part II forms the empirical heart of the book. In the next three chapters, I summarize the themes which emerged from the original research, which was the starting point for this book. The research involved in-depth conversations with senior, influential and successful stakeholders with a direct or indirect interest in the music industry.

According to their functional roles, the following stakeholder interests were notionally represented by the participants, though they were all keen to state that the views they expressed were their own, and not officially representative of their employers or clients. The job titles of the participants are listed in parentheses[1]:

- **the recording industry** (Chairman, IFPI; Former CEO of PolyGram and of EMI Music; EVP and CFO of Universal Music; Head of Digital, major record company; Independent Producer/A&R Executive)
- **the 'creatives'** (Artist Manager; Composer/Producer/Recording Artist)
- **radio broadcasters** (Head of Interactive Music, BBC; CEO Digital One, an independent commercial broadcaster)
- **music retail** (CEO, HMV)
- **new media music service providers** (SVP Marketing, MySpace; CEO, Independent Music Service Provider)
- **mobile phone operators and handset manufacturers** (Head of Strategy, Orange; Corporate Strategy Manager, Nokia)
- **investors** (New Media Financier)
- **government** (UK Minister for Intellectual Property)

It was not credible in the research conversations for me to adopt the identity of a wholly objective and independent academic researcher, even if such a thing exists. Some of the participants were acquaintances made through my previous roles in the entertainment industry.

Even those with whom I had no prior acquaintance became, through the process of engaging them as participants, aware of my background in the industry. Consequently, the conversations represented a different dynamic from the conventional researcher-participant inquiry. They are better described as business dialogues amongst peers, including the strategic musings of leaders within in a relatively safe and comfortable environment. As such they are a rare record of a more intimate and reflective industry discourse at a particular moment in time. Obviously there were still elements of self-promotion, power positioning, identity management, scepticism, mistrust, conformity with normative pressures and standardized forms of expression, and any number of other things which potentially interfere with the objective truth-seeking ideals of the traditional research interview. But, as power relations and identity constructs were part of the research focus, these things do not stand in the way of the research goals. Besides, such potentially distorting factors are a part of naturally occurring talk and if their existence is acknowledged from the outset, then at least one can then embark on analysis of the text without the pretence that the interview was undistorted by the interests and idiosyncrasies of the participants.

The conversations were semi-structured, with a common set of questions based on the scenario work developed at Royal Dutch Shell (Senge et al. 1994). This includes what they would have liked to have known ten years ago, what would they consider to be good and bad scenarios for the future of the music business, and what questions they would ask of an oracle. The participants' explanations and interpretations are presented with a high proportion of textual reference from the transcripts, and at this stage, with minimal interpretation other than to aid an understanding of industry terminology and context for the non-specialist reader.

I begin by identifying a more-or-less consensus view of the traditional *value chain* for recorded music as a provisionally fixed point from which to begin a journey of analysis. The most commonly cited challenges for the recorded music business are then introduced: the disruption caused by technology, and in particular, CD copying, peer-to-peer file-sharing, affordable and accessible domestic producing and recording technology, and the web as a medium for social networking and direct artist promotion. There is much consensus that these developments are attributable causes of change. However, it is the dissensus which is more revealing. Disagreement centres on whether they are positive or negative developments, and whether they should be encouraged or protected against. I am less interested in whether those negative and positive interpretations are logical or defensible accounts of 'reality' and more interested in how and why the interpretations are framed as positive or negative. The attribution of credit and blame, and of what is fair or unfair, right or wrong, and how these perceptions open up or close down opportunities for strategic action are the focus. Whether consciously or unconsciously deployed, it is these narrative techniques which are of most interest to anyone concerned with understanding the past and future evolution of the cultural industries.

Chapter 3, *Value Shift*, refers to changing perceptions of where the value in music consumption lies. It includes the impact of the digital commoditization of music and the apparent diminution in the value attributed by consumers to high-quality recorded sound and physical packaging, in favour of convenience and services which enable discovery, sharing and personalization. It explores the value propositions of competing stakeholders who have an interest in devaluing some product conceptions of music relative to their own consumer offerings, such as mobile phone companies and social networking sites. In the context of the 'democratizing' properties of new media and technology, I highlight the metaphor of slicing and baking pies to present the various arguments around the precise location of the core values provided by record companies to artists. In a wider societal context, the continuing relevance of the role of record companies as cultural intermediaries is questioned.

In Chapter 4, *Custodial Tensions*, the theme of custody is broken down into two subheadings: *cultural custody* and *economic custody*. Cultural custody includes the patronage of artists and the guardianship of the cultural legitimacy of art, fashion and taste, and the competing protection claimed for individual freedoms to enjoy, create and express. *Economic custody* groups together all the practical, ethical and social aspects of the debate around the protection of recorded music assets and the control of distribution channels. In particular, it explores how protection strategies, via technology and legal process, are constructed as both causes of, and solutions to, the industry's problems.

Chapter 5, *Hindsight*, concludes Part II of the book with some interpretations from participant hindsight reflections on what happened in the music industry in the previous ten years. The observations are considered in the context of Weick's (2001) concept of sense-making, and illustrations are given of how perceptions and openness to change can be desensitized by the kind of clear and public commitment to strategic choices and visions which industry actors feel compelled to make.

3
Value Shift

There is a common view of the value chain of the mature recorded music business as illustrated in Figure 3.1:

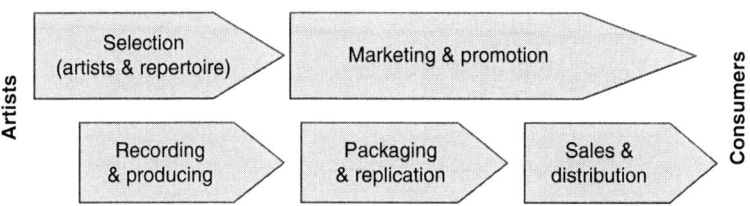

Figure 3.1   The value chain of recorded music

It is only one picture, a prevailing one but not uncontested, and establishes a provisionally fixed point from which to begin a journey of critical analysis. It constructs the music business as a series of value-adding activities, commencing with the artist and ending with the consumer, with the record company characterized as the dominant contributor of value and controller of many of the processes in between. The hard-to-define skills of artist and repertoire selection ('A&R') are held to be exclusive to the business, as is much of the mystique surrounding the marketing and promotion of music and artists. Descriptions of discovery, nurturing and development suggest a theme of patronage, whilst filtering and taste-making are more indicative of the role of cultural intermediary. In contrast to the intangible nature of these roles, there are parallel production and distribution processes, which are more tangible and which are described in mechanized terms more familiar to industry. This is the world of recorded music products in predominantly physical terms: to be replicated and disseminated to all corners of the globe in great volumes, and to be collected and valued by consumers as cultural artefacts embodied in technology. This value chain, which existed largely intact for the century following the invention of the phonograph, peaked in 1999.

Although participants refer to the existence of all elements of the chain, there is a notable variance of emphasis around which links are regarded as the core areas of expertise and value contribution. These differences are manifested as roles and identities. They become more sharply defined when the chain is described as fragmenting due to 'disintermediation' by new stakeholders. The resulting fragility of identities leads to some rich discursive material.

Parasitic activities

It is interesting to view the economic dominance of the traditional value chain against what I contentiously refer to as 'parasitic' activities. These are economic models which are not perceived to be directly part of the recorded music value chain, but which have indirectly derived the majority of their revenue from its existence. The term 'parasite' is contentious because the record companies have derived mutual benefit from these neighbours. They lived in mostly symbiotic equilibrium until the parasites came to be seen as outgrowing their dependency on their record industry 'host'. As the host revenues began to decline, the parasites continued to thrive. As new revenue streams appeared, they began to encroach on host territory (see Figure 3.2). Taking a more historical perspective, it could of course be argued that the recording industry is itself a business model which is parasitic on the activities of musicians, performers, composers and impresarios.

As barriers to entry have been falling, most of these businesses have been exploiting opportunities to enter the traditional record company chain of origination, promotion and distribution. Where the relationships fall out of equilibrium, the discursive properties of the conversation texts reveal their performative qualities, such as claims to authority and legitimacy, discrediting of other contenders, attribution of blame and implications of injustice.

Credit, blame and justice

Though there is much diversity in the perceived causes of the decline in record companies' fortunes, the focus of contention is on the

Business model	Traditional sources	New/growing sources
Consumer electronics	Manufacturers of home hi-fi, walkman	Manufacturers of iPod/mp3 players, 3/4G mobile devices
Radio	BBC, commercial radio stations	DAB, internet and satellite radio, subscription services
Music television	MTV	International talent show formats e.g. *X Factor, Idol* & *The Voice* franchises
Talent management	Fragmented, individual agency structure	Artist branding, corporate sponsorship, combining with concert promotion
Concert promotion	Fragmented impresarios and venue owners	Consolidating with talent management, e.g. AEG, Live Nation, Clear Channel
Merchandising	Cottage industry (parasite on concert promotion)	Expanding beyond 't-shirts', rights acquired by large entertainment corporations
Recommendation, discovery, sharing	Music journalism, fan clubs	LastFM, SoundCloud, MySpace, Songkick Yahoo Music, Twitter#Music, Orange Monkey
Synchronization (usage in TV ads; movies)	Mostly passive granting of rights on request	Pro-active pursuit of opportunities as part of sponsorship/branding growth
Publishing	Licensing and collecting; mechanicals (CDs) are primary revenue source	As mechanicals fall below 50%, growth through aggressive negotiation of new media rights and pro-active pursuit of commercial opportunities

Figure 3.2 'Parasite value chains'

anticipation of, and on the strategic and behavioural *reactions to,* developments in the environment. By contrast, with regard to the environmental developments themselves, there is broad consensus: the industry value chain was disrupted by several consistently articulated innovations in technology, media and design, as illustrated in Figure 3.3.

Though their origins date back to the late 1990s, these disruptions are new millennium phenomena in terms of their adoption by the mass market. Some other causes of problems for the industry are

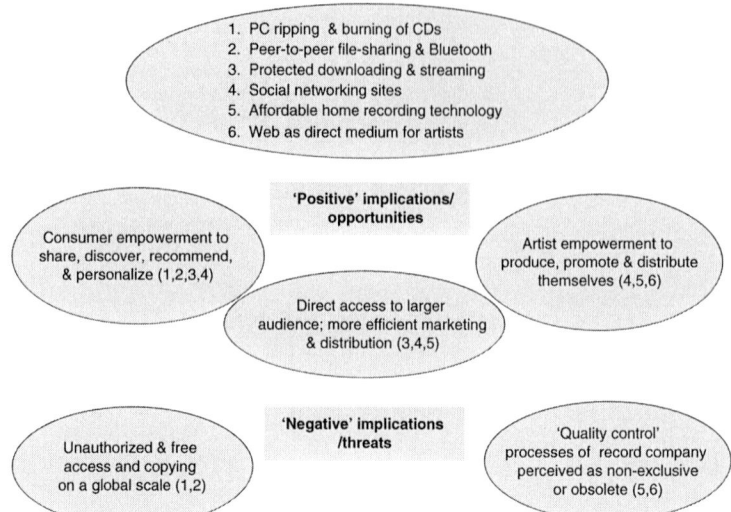

Figure 3.3   Technology disruption to the value chain

less emphatically cited by participants, such as product devaluation through aggressive supermarket pricing, free CD give-aways in newspapers and magazine cover-mounts, and growing competition for consumers' disposable income from PCs, videogames and DVDs. However, such trends mostly arose many years earlier than the developments listed in Figure 3.3 and did not prevent the industry from continuing to grow to its 1999 peak.

Given the consensus around the primary value-chain disrupters, the main thrust of the analysis follows the participant texts in exploring the more discursively interesting dissensus, which centres on the positive and negative implications of the disruption. In conventional strategic discourse, these polarities would be framed as opportunities and threats from the environment; and the strengths and weaknesses of the record companies in their ability to deal with them would be the primary area of interest. My categorization of positive and negative in Figure 3.3 is already questionable, and entirely dependent upon whether one gives primacy to the rights and interests of artists, consumers or record companies. I therefore do not try to establish any kind of objective conclusion on the merits of these interpretations, as I am more concerned with how and why they are framed as positive or negative, and whether blame or credit is being attributed.

With that in mind, the content under the next few headings relates to the perception of the value of music and how it is changing. They provide a framework which captures the participants' sense-making of what has happened and what might yet happen in the business of music.

Value consensus

> I actually wouldn't ask the oracle the question: 'will people still love music?' because I have no doubt that that will be the case. You know, the question I would want answered is: 'what would consumers value about the way in which they access and consume music?'
>
> (Former CEO, HMV)

This quote is representative of a general confidence about the permanent and deep-rooted value of music in a cultural, if not economic, sense, and this is consistently shared by all participants. Music is variously described by them as: *more popular than it's ever been*; *one of life's eternal values*; *too important to the world*; and *part of people's lives*. Though the overall value attributable to the music experience is perceived to be stable, there are some elements which have lost value and others which have gained. The net economic effect of these changes for the *recorded* music industry is negative, implying that the elements of devaluation are core to the existing business, whilst the new areas of growing values have been difficult for old companies to embrace and control, economically speaking. Therefore, there is a common feeling amongst participants that values have shifted detrimentally for the industry: *it used to be something special* but now *it's lost its value*; *you're not quite sure what you're paying for*; young people *think you don't have to pay for music*. The remainder of this chapter is split between those areas of the business which are seen to have lost value, and those where new value is being created.

Diminution in value

I have grouped the discussions relating to diminution in value into four headings:

- digital commoditization
- physical-visual
- physical-aural
- the honey trap

Digital commoditization

Digital commoditization is a reduction in the scarcity value of music, resulting primarily from its decontextualized proliferation and widespread availability through unauthorized channels (copying and sharing):

> You're just delivering a set of information, bits... in terms of music. If it's commoditized in terms of it being available from other sources, it's not that interesting. One of the data points that I saw a year or two ago was that within 5 years' time it'll cost about $200 to buy enough hard drives to store all of the recorded music made in the US after 1950.
>
> (Corporate strategy manager, Nokia)

The retail cost of filling the largest iPod (160Gb) with approximately 38,000 purchased songs would be around £30,000. The device itself costs less than £200. When such vast amounts of desirable content can be stored on, and easily shared between, small devices, it is no wonder that there is a dilution in economic value of music; it is also no wonder that the iPod has been so successful.[2] Of course most iPods are far from full, but even so, statistics indicate that the *average* iPod of 18–24-year-olds contains 842 unauthorized songs, representing half of the songs on their iPod.[3] The retail value of 842 songs is £665, and if this were extrapolated to all the iPods and MP3 players in use, its value would equate to more than ten years-worth of global recorded music sales.

There appears to be no public consensus on whether or not the prevailing retail price for digital music represents good value, but there was consensus amongst participants:

> I mean if you subscribe to Napster [*authorized service*] you get pretty much everything you want for £10 a month, which is a pint of beer a week, so it's not actually really... That's effectively almost free to be honest with you.
>
> (Head of interactive music, BBC)

> I'll go on iTunes in the next couple of days and I'll pay £1.60 to have two classic tracks, two just fantastic tracks for £1.60, which you know, subject to any problems I'll keep forever. It's just unbelievable value.
>
> (Chairman, IFPI)

To his great frustration, the view of the chairman of the IFPI is not shared by his teenage daughter:

I'm stunned. 'You have no problem paying £30 out of your allowance, a huge percentage of it, for a mobile phone?' No problem. It was a 'must have'. No problem, not thinking about it at all, absolutely had to have it. So completely stunned. So I thought I'd examine this. I said to her, 'I know you love music, so what if, for £40 instead of £30, you could have all the music you want, everything you want included in your £10?' She said 'no'. I said, 'why no?' She said, 'it just doesn't feel right'. So I said, '£8 then?' No. '£5?', hesitated, but still no. I wasn't going to depress myself by going below that.

Not 'feeling right' is a disturbing comment. It suggests that young consumers don't have a clear perception, or have a very different perception, of the economic value of music.

now there's music everywhere [...] you can see people are engaging in it but there's almost I think... I'm sounding like that grumpy old man programme, but there's almost too much of it. So it's lost its value

(Composer/producer)

How do I get the public to understand the value of music? Because they do love music.

(Chairman, IFPI)

They said: 'We think you (*radio*) can help us (*record companies*) in communicating to the punters about how to value music, how it should be positioned, about new music.'

(CEO, commercial digital radio)

These comments presume that the public need educating about the value of music. A very interesting test of consumer value occurred in late 2007, when the band Radiohead chose to make their new album *In Rainbows* available on their website, prior to it being released on CD, with the invitation that consumers could pay whatever they liked. The experiment attracted much media attention, demonstrating quite polarized views, both in regard to the bands intentions and to the interpretation of its success or failure. It was reported that approximately one-third of people paid something, and of those who did pay, the average paid was around £3. Despite being widely reported as being available for free, there was actually a 45 pence transaction charge from the website, meaning that credit card details had to be given. On the day of its release, 400,000 illegal file-shares (i.e. not via the authorized site) were made and this number had grown to 2.3 million after three weeks,[4] despite being available to people in an authorized way directly from the artist's site. The implication is that either 45p is considered too much, or that file-sharing consumers are by habit tied to their usual sources, and that any extra administrative effort is a deterrent from making an authorized purchase.

Some have argued that the experiment was a failure. Paul McGuinness, manager of the band U2, described it as having 'back-fired'. Meanwhile U2's Bono described it as 'courageous and imaginative in trying to figure out some new relationship with their audience'.[5] Others have argued that it was a 'success, contributing to the album topping the charts in both the UK and United States and a wildly successful worldwide tour'. They add 'when it comes to judging whether an album is a success these days, the old metrics just don't cut it'.[6] By comparison, my research participants were mostly sceptical about the experiment:

> What's our album worth? Pay us what you think? Well it's all very well for these guys because they've already made their millions. I find it annoying when you've got people like Radiohead saying, pay me what you think the album's worth and all this. It's almost like a little...actually they're getting great publicity out of it. [...] I'm quite surprised with the numbers who paid for it. They've done quite well actually.
>
> (Composer/producer)

> we had the Radiohead experiment, which was a very interesting experiment, very commendable, very innovative, all of those things. The terribly sad thing was that so many of their fans and so many of the consumers still chose to steal it, for want of a better word. And that's really sad. When an artist had come out and said, okay, look there's a cost of a credit card, a cost of 45p, beyond that you can pay what you want. [...] What that shows you is that free is an unbelievably compelling proposition and it's unbelievably price sensitive, because at 45 pence for an album it's too expensive if you're competing with free. At different times you read press saying, if only record companies didn't overprice things there wouldn't be a problem. What this showed was that's not really the issue. It's free. It's the availability of free. And if you can address the availability of free, then you have a different dynamic
>
> (Chairman, IFPI)

If the biggest competition is coming from 'free', the challenge for the industry is therefore to provide products, services and experiences which are better than what is freely available.

Physical-visual

The second element of value decline is physical and, to differentiate it from the next category, I will call it physical-visual. The most tangible loss of value in the digital age is the physical packaging of music:

> kids are not going to get the same buzz out of buying an album and seeing a great cover of the Santana album or something like that, and all the notes and the whole vibe of it and then going round to your mate's house and taking all your records with you and sitting there and everyone's playing records and the excitement of it...you'd read the cover and you'd look at it and it was like a piece of art and I remember then every record company had their own art department all coming up with great designs, and to me that was a whole stream of income that just went...went down the plughole, because people were collecting that. It was something you could get hold of. It was tactile.
>
> (Composer/producer)

Actually the value of physical packaging lies not just in its aesthetic aspect, but also in its informational aspect through the detailed notes which, pre-internet, was one of very few places one could get more information about the artists and the songs.

> I want to see all the notes. I want to see who played them. I want to know who produced it. I want to know all the musicians. I want as much information as I would have had on a 12' album. You don't get it. You don't get anything. So then you just have to go on Google and use that, right, who produced it and you have to go about it this way. To me it's insane. If you're going to pay that sort of money, at least give you your money's worth.
>
> (Composer/producer)

The love of the physicality of the music product tends to be a nostalgic concern of the older participants, and, according to this participant, this value had already mostly been lost when vinyl LPs were replaced by *this bit of non-collectable piece of crap plastic* [the CD] *which you might as well put in a brown paper bag because you can't see anything.* Nevertheless, others do see continuing value in the CD package:

> People do still want beautifully designed proper things of artists they like... very very informative booklets. They don't mind paying extra for that [...] It's great that Radiohead have done a £40 box set because that's what the fans want.
>
> (Independent producer/A&R)

Physical-aural

The third element of value loss is physical-aural. Improvements to recording and playback sound quality are generally acknowledged to have been a primary driver of growth in the recorded music industry from its very beginnings over a century ago. Up to and including the launch of the CD format, the marketing of both content and equipment had persuaded consumers of the necessity of having the finest quality hi-fi sound. Yet in less than a decade, this hype seems to have largely evaporated. There is relatively little concern around the consensus view that the most popular compressed digital file formats such as MP3 and iTunes' AAC are notably lower than CD quality:

> I think it's a few of us have cared about sound quality, but when I was a kid I had a medium wave radio. That was fine for me. It's probably the same as that. Singles never sounded that great, albums didn't. I mean... I don't like the sound of MP3s personally, but other people don't give a monkey's. They just want to hear the song.
>
> (Independent producer/A&R)

> the record industry maybe thought, 'well you know, Christ, we sold it [*the CD format*] on higher quality. How can we possibly go back and say, if we compressed it to a tenth of the size of its original thing it will be fine?'
>
> (Music service provider)

Reduction to 'merely acceptable' quality sound has seemingly been a small sacrifice for the consumer compared with the benefits of storage capacity, portability and 'share-ability'. In any event, the industry has been spared potential criticism for previously overselling high-quality sound to those who either couldn't detect or didn't care about it. But as CD sales continue to fall, the true cost of relinquishing 'hi-fi' as a marketing angle may ultimately be measured in the devaluation of recorded music altogether.

Honey trap

The fourth and final category of music devaluation I've called 'honey trap'. This metaphor refers to the exploitation of music in order to sell other things. The best example is the iPod:

> so Steve Jobs has come to market and he has invented an incredibly clever device and he uses something called iTunes as a store, basically as a honey trap for people to... you know, to buy the iPod.
>
> (EVP/CFO Universal)

> You can see Apple, growth of profits, billions of pounds out of an iPod, which wouldn't do anything about music. How sad is that? You know, it's wrong actually. And that's where I have a sympathy with the recorded music. They've made billions and billions out of this stupid little device that would have done nothing without the content.
>
> (Digital music service provider)

> Music is a diminishing component, but you maybe use it for selling small electronic devices
>
> (Former CEO, HMV)

Despite Apple's claims about the billions of songs it has sold through iTunes, they represent only 3% of the songs carried on iPods (Jobs 2007). Similarly, Apple's profits from the songs are only a tiny fraction of the profits from the sales of the iPods. There is no doubt therefore that, in economic terms, Apple is a consumer electronics manufacturer more than it is a music retailer, and consequently has every interest in keeping the levels and structure of music pricing as attractive to the consumer as possible.

But the honey-trap metaphor applies to many other channels of music distribution, such as heavy discounting by supermarkets (... *it must be almost like a loss-leader for them isn't it?* Head of interactive music, BBC), and most notably the mobile phone industry:

> I think the telcos view music as nothing other than a honey trap to get their customers
>
> (EVP/CFO Universal)

Music is a big selling point to them, I mean that whole Vodafone offer this Christmas with MusicStation, that was basically Vodafone's only piece of marketing for Christmas was Music Station.

(New media financier)

And maybe artists feel that they're being undervalued and there's a level of arrogance there. I am sure there is. But you know, if you said to Justin Timberlake, you know, your tracks are fairly worthless but what they're pretty good at doing is selling more mobile phones for Nokia, I am pretty sure he wouldn't necessarily like the sound of that...[...] I think music is a very powerful way to...I suppose the question is, I'm trying to answer my own question, but is it a...is it now just an add-on, just something that's given away in order to get something else which is of more value.

(Head of strategy, Orange)

There were many more examples of music being referred to as a marketing tool for other things. Part III of the book will further consider this strategic value of music, in particular to the relationship between the music and mobile phone industries, but for now I will move on to new value opportunities.

New value opportunities

Again, for ease of reference, I have grouped the discussions relating to new value opportunities into four headings:

- live music (including the 360 degree model)
- the subscription model (including flat-fee licensing)
- the advertising-funded model
- integrated mobile phone services

Live music

One of the more interesting phenomena commented on by many of the participants was the opposite fortunes of the recorded and live music businesses in the new millennium. In 1999, the UK consumer spending on recorded music recorded music was more than four times that of live music. Ten years later, live music had increased more than threefold, whilst recorded music fell, with the result that live music now generates more revenue than recorded music[7]:

It's amazing. Actually it's a real phenomenon. It really is. And I keep wondering whether it's a bubble that's about to burst and people are going to go, 'sod it', you know, '£60 for a ticket which is going to last an hour and a half and I don't have anything to take home with me?' But what'll happen is, you know, there will be models where you go to the gig, you pay your £60 and at the end of the gig you get a copy of the gig to take home so it's more of an attractive proposition... didn't Prince give a copy of the album away with the gig ticket price?

(Head of interactive music, BBC)

We were talking about paying for music, and one of them [*16-year-old in a focus group*] said, 'why would I pay to listen to an advert? I wouldn't pay to watch an advert for a car, so why would I pay to listen to an advert for a band?' And they all started nodding furiously. And the perception is that the recorded performance is the advert for you to buy the t-shirt, or go and see them live. And that's part of the value chain. [...] The world of paying just for straight content is over.

(Head of digital, major record company)

The only thing that has got any special quality now is to actually go and see a live thing. Which is fascinating when you think there was a time, not too long ago when, if you could put a tour together you were probably going to lose money. It wasn't that long ago that bands played in pubs and <u>paid</u> to play. [...] That makes it even funnier still, you know. It's pure promotional. Albums are purely promotional now for a lot of these acts and their merchandising and live gigs... so the music's become an add-on.

(Composer/producer)

I think people are so fed up with electronic entertainment, computer screens and all the rest of it they just want the taste of the real thing, I think, you know... the moment. I think there is an excitement through seeing something live...

(Independent producer/A&R)

The growth in live music is not necessarily indicative of the commercial prospects for *new* artists. The vast majority of live revenue growth has been from 'superstar' artists who were already well-established before 1999 and who can generate in excess of $100 million from a concert tour. Of the top 40 highest-ever grossing concert tours,[8] though 90% of them occurred post 1999, only one (Lady Gaga) is from an artist who has come to prominence in the new millennium. As far as new and developing artists are concerned, record companies often regard tour support as one of their significant items of promotional cost. This indicates that, for most, live revenues are not sufficient to cover the costs of touring, and that the promotional benefit of a tour only makes sense in the context of building a longer-term career or 'brand' for the artist. The growth in live performance at the expense of CD sales in aggregate therefore seems ironic, but is underlined by the growing number of both older and younger artists (e.g. Prince, McFly) who have chosen to give their CD away free for promotional purposes, underlining the point that the acquisition of the physical album no longer has scarcity value, and that the live experience is a richer source of that *special quality*, or *the real thing*.

The 360 degree model

> Sanctuary Group … is perhaps the model for the music company of the future, with 360-degree participation of all the related revenue streams meaningful to artists … This strategy is paying off and positions them as a potential long-term survivor. They are filling the vacuum left by the major labels with a more enlightened business model, based on a philosophy of artist management.
>
> (Kusek and Leonhard 2005, p. 113)

The diverging fortunes of live and recorded music have led to much discussion around the so-called '360 degree' deal, whereby record companies extend their contractual rights with artists to include a share of their live performance, merchandising and all other revenues:

> It seems to me as though, if the record industry promote an artist, it's not unreasonable for them to get some part of the concert revenue, merchandising revenue … what's its name called? The 360 type thing
>
> (Digital music service provider)

The record companies argue that they can no longer afford their previous generosity and the deal must change because the marketplace has changed. Though some people find this logical and fair, it is fundamentally changing the traditional terms of revenue division between artists and record companies, often without the record companies being in a position to provide an extra service:

I think our competition [other record companies] have a different view which is...you still write the big marketing cheques. Then in order to help finance those marketing dollars you are requesting that your artists give you a percentage of their earnings from other activities. I think that's wrong, because I think if you ask somebody for a participation for something where you do not provide a service, I think they view that as stealing.

(EVP/CFO Universal Music)

The indies [independent record companies] are no better placed to become 360 degree models than the majors.

(Artist manager)

Sanctuary, an independent music company and pioneer of the 360 degree model, ran into financial difficulties, and in 2007 was bought by Universal Music:

By snapping up Sanctuary, Universal has now got its hands on a whole new set of artist services, effectively grabbing a bigger slice of the music pie. Record companies 'never traditionally had a slice of all of the cash an artist made – just a share of their recording revenue', says Andy Gemmel of Big Print Music. 'The label generally created the artist, therefore that's a bit unfair.' Industry talk suggests that Universal's takeover of Sanctuary will force rival majors EMI, Warner and Sony BMG to actively seek out new revenue streams.

(BBC News 13 September 2007)

Just as record companies are trying to extend their rights and their businesses to incorporate live and other revenues, so the live music promoters and corporate sponsors are trying to become record companies. Live Nation is a very large concert promotion company which has been in the media spotlight since 2007 due to its strategy extending its acquisition of concert and merchandising rights by acquiring the recording rights of some superstar artists:

A fierce battle has broken out among top executives at Live Nation Inc. over the concert-promotion company's ambitious strategy to reshape the struggling music industry by making wide-ranging but expensive deals with artists such as Madonna and Jay-Z. The battle is over the limits of that strategy, in which Live Nation has pledged hundreds of millions of dollars to a handful of performers in return for exclusive rights to release their recordings, promote their concert tours and sell T-shirts and other merchandise bearing their images.

(*Wall Street Journal* 12 June 2008)

With some artists entering into corporate brand relationships with fashion, cosmetics, drinks, phone and movie companies, the Live Nation strategy of owning the complete brand is a very interesting development of the past five years, and one which raises questions of cultural-industrial identity and the right to participate in the value chain of music, and the associated obstacles to collaboration and survival.

The live music business is in many ways the easiest to talk about as an opportunity for record companies because it is highly visible, generates clear revenue streams and is so closely related to their existing business. It is however a mature business which, despite its recent growth, is unlikely to be able to satisfy all of the stakeholders competing for a slice of the music 'pie'. The pie metaphor is a common one in strategic discourse. It refers to the way in which economic value in a deal, business activity or market, is shared amongst participants. A relevant example of its usage can be seen in a recent blog which refers to a meeting of US music industry lawyers to discuss the 360 degree model:

Fred [*an artist lawyer*] asked Rand [*a lawyer at a US record label*] if Interscope was a pie-slicer or a pie-baker.

(Lefsetz 2008)

The Lefsetz letter is a well-known blog commenting on the US music industry, and one research participant (the artist manager) was of the view that it is *read by all the heads of record companies*. I quote it here because I think that the metaphorical distinction between pie-slicer and pie-baker is relevant. In pursuing the other (non-recorded music) revenue streams of artists, record companies are focusing on existing, mostly traditional economic models. In this sense, they are re-slicing the pie to give themselves a bigger share. Fred's question is important because it implies that record companies could bake new pies; that is to say, find new business models which will generate *incremental* economic value from music, rather than resorting to taking *existing* money from the pockets of their artists. This challenge is acknowledged by one record industry participant in response to getting into the merchandising business:

I think the whole thinking here has to be to grow the size of the pie, otherwise there's not any real point in doing it.

(EVP/CFO Universal Music)

Baking new pies

Pie-slicing could also be described as a being synonymous with a *zero-sum game*, where the winner's gain is exactly equal to the loser's loss. The latter term is used by the corporate strategy manager at Nokia to describe the relationship between the music and the mobile phone businesses, and in particular his view that no new value is being created by the collaboration. There is an enormous amount of public discourse on this subject of how to create new value, or in Fred's words, bake new pies. I will start with the solution which most participants agree represents the best economic model for the future of music, the subscription model.

The subscription model

I think that model will win and I don't know if it's 5 years or 10 but I think it will ultimately be the way that we consume music will be by subscription. And I think it'll be mainly by mobile subscription.

(New media financier)

things that people particularly need to be looked at are things like the subscription models which seem to be almost the most talked about future model.

(Artist manager)

Paid music subscription services, well I'm going to be optimistic and say, if the record labels get it right, then actually they could be a clear winner.

(Head of interactive music, BBC)

where I think there is optimism for the future, I do believe that the future of the music industry will come back... the good fortunes will come back through a subscription model. That's where I think it will end up.

(Chairman, IFPI)

I've always been a believer in subscription...

(EVP/CFO Universal Music)

The basic premise of the subscription model is that, for a periodic fee (e.g. £10 per month), consumers have access to a complete library of all the music ever made digitally available by record companies. As long as they keep paying, the access remains; once they stop, the access is closed. If the only users to take subscriptions were users who were previously buying two CDs per month, and who then stopped buying CDs as a result of the subscription, the model wouldn't generate incremental value. But the economically attractive thing about subscriptions is the hope that they can become, like cable or satellite TV, a default lifestyle necessity, a 'utility', a 'must-have', and a non-discretionary cost in the household budget. In aggregate, more people will pay for subscriptions than were actively buying music before, and many will continue to buy music (CDs or downloads) on top of their subscription. Most subscriptions (when adopted on a mass scale) will in effect be under-utilized. As a result, there is incremental growth in the music market and a new, or bigger, pie is baked. Or so the theory goes. In practice, subscription models are still struggling to find mainstream adoption to the same extent as TV.

Picking up the domestic utility quote above, another participant makes reference to music being like water:

> If music is much more available and accessible and there's sort of a ... division of music like water ... Gerd Leonhard and others ... make it much more accessible then there's just going to be much more activity related to it.
>
> (Corporate strategy manager, Nokia)

Music-like-water became a widely used term following the publication of Kusek and Leonhard's (2005) polemic, *The Future of Music: Manifesto for the Digital Music Revolution*, whose opening chapter is entitled 'Music like Water', though they attribute the analogy to David Bowie in 2002. It is a variant of the subscription model, and as an argument it is very convincing. The authors, and other anti-recorded music industry writers, claim that the industry is vehemently and ideologically opposed to the music-as-water proposition, 'fighting it tooth and nail' (p. 8). But, as evidenced by the participant quotes above, the industry is, and has, for many years, been convinced by a utility-type model. Kusek and Leonhard's claim seems either misinformed (being confused with the industry's anti-file-sharing efforts) or deliberately misleading, in order to suit their overall argument that record companies are closed to any new models.

The most frequent attribution of failure of subscriptions was that, at the time of this conversation, they had not successfully incorporated a solution which facilitates portability, i.e. one that integrates with an iPod, MP3 player or phone. Consequently, the music service could only really be enjoyed from a computer. This constraint had less to do with technological capability than with the collaborative challenges between device makers, software providers, entertainment companies and media providers referred to above.

Although all parties had much to gain and to lose in such a negotiation, the music industry was not necessarily the one most at risk, and there is a view that the main obstacle to the subscription model was Steve Jobs himself:

> **Researcher (JW):** if Steve Jobs had chosen last week to announce, rather than movie rentals, a subscription model on music...
>
> **Head of digital, major record company:** Suddenly the world would change.

Jobs was always clear, at least publicly, on his anti-subscription view as this extract from a *Rolling Stone* magazine interview illustrates:

> they're [record companies] fairly vulnerable to people telling them technical solutions will work, when they won't [...] because of their technological innocence. [...] At first we [Apple] said: None of this technology that you're talking about's gonna work. We have Ph.D.s here that know the stuff cold [...] We said: These [music subscription] services that are out there now are going to fail. [...] Here's why: People don't want to buy their music as a subscription. They bought 45s; then they bought LPs; then they bought cassettes; then they bought 8-tracks; then they bought CDs. They're going to want to buy downloads. People want to own their music. You don't want to rent your music – and then, one day, if you stop paying, all your music goes away. [...] It's cheaper to buy, and that's what they're gonna want to do. They [record companies] didn't see it that way. There were people running around saying: No, we want a subscription business. We [Apple] said: It ain't gonna work.
>
> (Goodell 2003)

Jobs makes a swift leap in this interview from a technological argument to a consumer-adoption argument, and on both counts he is dogmatic in his pronouncement of doom on the subscription service. Ironically, 11 years later, recognizing the success of services such as Spotify and Deezer, and the entry of Google into the music subscription market in 2013 with its Google Play Music *All Access* service, Apple finally took the plunge into the music subscription market by acquiring Beats

Electronics, stating with confidence that they will be the first ones to get the subscription service right.[9]

It may be fruitless to dwell too long on the *rationality* of the question of whether consumers want to buy music, rent it, or have it supplied as a utility:

> there seems to be some piece of consumer mentality around, anything I get online is free. Now whether... and I certainly think that's the problem with a 'pay per view' or 'pay per access' type model. I think, up to a point bundled content or services under the form of subscription may be easier to sell to people. You know, maybe we will view music a bit like a utility where you pay a sort of annual fee and you have a sort of 'all you can eat' type approach and as you know, if you look at the size of the market, actually you wouldn't need to get that many people to pay that much money for that to be quite viable. [...] If I was behaving rationally a rental model would be much more in my self interest rather than an ownership model. But I don't think consumers think and act that way.
>
> (Former CEO, HMV)

The subscription model has been very slow to take off, but in 2014, with the clear endorsement of Apple and Google, its time may have finally arrived. There *are* still significant obstacles, to which I will return in Part IV.

Meanwhile, a variation on the subscription model emerged over the last decade which had similar cultural and emotional obstacles to its development. It is called flat-fee licensing.

Flat-fee licensing

The 'music like water' analogy leads one step further to another way of trying to resolve the problems of the music industry:

> The only business model I see that has any success... I'm a big fan of...Jim Griffin. I don't know if you've come across him. Industry pundit. He has been spending a lot of time around this flat fee licensing.
>
> (Corporate strategy manager, Nokia)

Jim Griffin describes himself as 'dedicated to the future of music and entertainment delivery systems, and works as a consultant to absorb uncertainty about the digital delivery of art' (Griffin 2008). He attracted much media attention in 2008 through his work on brokering a deal between internet service providers (ISPs) and music companies. His proposal, as reported, was to collect music fees from consumers via their ISP bills, essentially by charging a single flat fee to all internet users, irrespective of whether or not they are music consumers. It may even have become invisible to the consumer by becoming wrapped up or hidden in the ISP's connection charge, at which point it becomes simply a negotiation between ISPs and music companies over what level of compensation they require for unrestricted licensing of music content. This might sound radical, but not if one thinks of the radio model (outside the US), where broadcasters pay a share of their revenues (e.g. advertising) to the record companies for a similar licence, and the consumer is unaware of the cost. Griffin's idea is commonly associated with Harvard Law School professor William Fisher's (2004) book *Promises to Keep*, which takes the flat-fee licensing idea to its logical conclusion by proposing the complete removal of copyright, to be replaced by a government-administered reward system involving free access for consumers, though funded through general taxation.

Griffin's proposal (or something like it) is just one of the potential solutions which the UK government had in mind in when brokering a memorandum of understanding between the ISPs and music companies five years ago. The government intended that these negotiations would result in a bigger pie through protection of 'the growing success story that is Britain's creative economy' (BERR and DCMS 2008). The upside is based on the same thinking as the subscription theory described above, and the success of a ubiquitous offering would be highly dependent upon ISPs, who inevitably wanted economic terms from the music industry which would make it worth their while. They were risky and ultimately unsuccessful negotiations because public acceptability and consequent consumer behaviour patterns were unpredictable. Griffin's proposal was described as 'an ISP tax', an 'extortion scheme' and a 'piracy surcharge', indicating the emotive language used in the debate. The dilemma is illustrated by a research participant who at the time of the interview was the government's first minister explicitly responsible for intellectual property. On the one hand he believes that the radio model of paying over a share of advertising revenues might be worth exploring:

> Most of the internet service providers and certainly the search engines now are very heavily reliant on big advertising revenues. There may be an appropriate levy against that as an income stream.

However, there is a distinction being glossed over here between network service provider, and provider of other gateway services associated with ISPs. If the solution is that ISPs include a music service within their monthly subscription charge, the principle becomes similar to a copying device levy. Yet earlier on in the interview, he gives his opinion, as a minister and as a consumer, on the unfairness of charging for things that people don't use, in the context of another proposal to charge a copying levy on digital devices:

> it would be pretty wildly unpopular here because there would be a number of people, and I actually, you know in a personal sense might be one of them, who doesn't use the fact that you can connect up bits of your hi-fi system to shift format. I use the different bits for the different things that they originally did and not out of fear of law actually. It's just as it happens how I use it. And generally speaking I don't think you can charge people for something they may never do. Somebody said to me the other day, 'but I pay my local council tax and I don't use all of the services'. And I said, 'the reality is that I'm making a big contribution through my council tax to schools. I haven't got any children in schools but I understand the importance of education in general for the benefit of everybody including me. If you want anybody qualified to do any of the jobs that I might rely on that's the way you get it.' It's not the same as playing a piece of music and nobody will ever see it as the same.
>
> (Government Minister for IP)

The difference in political acceptability between the advertising levy and the device levy seems to be primarily one of remoteness and/or visibility of cost, and this leads into the next area of new value opportunity, the advertising-supported model.

The advertising-supported model

Economically speaking, the subscription model is easy for business strategists to embrace because it is already a proven model in other media, such as TV and print. Another proven media revenue model is advertising. In the two most influential music markets, the UK and the US, it's easy to forget radio as an important advertising revenue source, despite the fact that radio is a primary mode of promotion and consumption of music. In the UK, this may be because the BBC operates on public licence income without advertising, and in the US because radio stations do not have to pay anything to record companies for the use of their recordings. In most other parts of the world however (including UK commercial radio), radio stations pay a proportion of their advertising revenues to record companies.

Referred to as the 'ad-supported' model, or sometimes the 'free' model, reflecting that it is free to consumers, record companies have been experimenting with licensing music and videos to internet sites for some years. Compensation is generally in the form of share of advertising or other service provider revenues, with guaranteed minimums. The music is streamed, meaning that consumers cannot obtain permanent copies on their computers. The offering to the consumer is marketed on some combination of discovery, recommendation, sharing and social networking, and, from the record companies' point of view, the thinking is that the promotional value of this usage is greater than the value of potential lost sales. Many internet businesses were launched on the promise of advertising revenues which have subsequently failed to materialize on the scale projected. One participant explained that when he was at Yahoo there were 60 million music streams per month. This sounds like an impressive number, but he went on to explain that with a CPM (tariff) of $50, and a click-through of 0.3%, this converts to advertising revenue of only £5,000 per month, which is less than one penny per song streamed.

Other popular examples of ad-supported music referred to by participants are MySpace and LastFM, though their revenues are modest:

> MySpace music, whilst it has a great importance in music, most of what people are doing on MySpace is social networking and then kind of consuming content second or taking content and kind of making it their own and creating their identities online with you know, music and video and adding it to like a richer profile experience. There isn't a whole lot of revenue there yet for anybody. [...] the expectation of free is alive and well and definitely isn't going anywhere. I think that that's never going to change anymore.
>
> (SVP Marketing, MySpace)

> I've been speaking with Michael Breidenbruecker [*LastFM founder*] quite a lot about LastFM and the stuff he's doing now, and he I think completely gets it as a concept, because music recommendations is just one element of life recommendations, which I think could be a big growth area.
>
> (Corporate strategy manager, Nokia)

Until it was overtaken by Facebook in 2008, MySpace was the most visited social networking site in the world and briefly had a very significant influence of new artist development. Expectations for the role and strategic value of music-related services were huge. Both MySpace (News Corporation) and LastFM (CBS) were purchased by global media companies for the strategic and synergistic value of their millions of users, rather than their foreseeable cash-flow. News Corporation failed to develop a business model commensurate with its 2005 $580 million investment in MySpace and lost out in functionality and design to Facebook. Despite it being valued at $12 billion at its peak,[10] MySpace was eventually sold for a mere $35 million[11] in 2011, illustrating the extraordinary high stakes for those competing in the arena of social media businesses.

LastFM has fared slightly better but still struggles with the ad-supported model:

> CBS executives quoted saying, 'well, this is of course our plan when we bought LastFM', so these are CBS people working in television and all they're thinking is, 'music? – nobody pays for it. We don't need to pay for it. Great. Excellent. How can we get people to come to ours? Oh, I know. Make it on demand. Put some advertising around it. Like telly, yeah, we know about telly, so let's just do that.' So they go the same way. So they end up doing the same thing. To be honest I'm really not claiming to be a genius because that's all that would occur to me is that you stream it for free and you ad-support it but it's not terribly radical.
>
> (Head of digital, major record company)

In spite of some sporadic excitement about ad-supported models, the music industry is treating it as just another media channel with a primarily promotional value in the hope that, once hooked either by a song or by a service, consumers will ultimately pay something for it. The subscription services Spotify and Deezer seem to have exploited this theory well, with clear migration between their ad-supported and paid premium services. Leading US internet radio provider Pandora has fared less well in converting its active users into paying subscribers, though it does have very impressive advertising revenues. The Nokia corporate strategy manager saw the ad-supported model as having much greater potential in a mobile phone context, which takes us to the fourth category of new value opportunity: integrated mobile phone services.

Integrated mobile phone services

> I think there's a huge future in digital lifestyle management, DLM. [...] the person then who would be choosing my music for my

holiday or my party or... I'd love to be able to go to Liverpool and have my phone take me on a musical walkabout and play the music each time I go past a venue that is relevant to the musical legacy in some way.

(Corporate strategy manager, Nokia)

His vision, while not directly an ad-supported one, is one where music can be incorporated with other commercialized services relevant to the mobile consumer and be used to support the long-awaited promise of the third generation (3G) experience:

To me the value-adding services are going to be around things like context and location and having the extra ability for me to get relevance out of the service that you're offering. [...] it [*the business model*] can't be based on trying to monetize bits which are extremely difficult to protect against. It has to be around layering at a higher level of value and the higher level of value is things that are much more personal and much more immediate and much more relevant. And so that's why I think the mobile device becomes a real tool for the music industry to go: okay we can actually start to get much more value. We can start to help improve the experience of music because we can't be based on just delivering the bits and expecting the same amount of money.

(Corporate strategy manager, Nokia)

This value-shift section ends with mobile phones because, as illustrated earlier by the daughter of the Chairman of the IFPI, this is where the consumer seems most comfortable spending money. Mobile phone companies have a number of attributes which lend themselves to music distribution. One is the ability of handsets to compete as digital music players, and a second is the immediacy with which consumers can satisfy spontaneous purchase decisions 'on the go'. This is particularly notable amongst children:

there's almost a level of showing off that people... you know, I've got this track before you

(Head of strategy, Orange)

They're able to pass it on to their friends quicker than they've ever been able to and I think capturing that excitement is a great thing.

(Artist manager)

It's absolutely immediate. It's in the school ground, you know, it's right there

(New media financier)

A third attribute is the ease and efficiency of the billing system, and a fourth is that mobile networks are tightly controlled, preventing unauthorized copying. In addition, as they expand into internet broadband services (as Orange has done), mobile phone companies can provide integrated mobile and online services. An example of this in music is the dual download which gives consumers the song on both their phones and their PCs simultaneously.

The early prospects looked good for music on mobile phones when the market for personalized ringtunes grew rapidly to over $4 billion in 2004, in spite of the fact that these 30 second clips cost up to three times more than their full-track equivalent. This appears to have been a transient phenomenon, and the ringtune market is now in decline. The £22.5 billion auction in 2000 of the UK 3G spectrum was based on the assumption that mobile phone consumers were prepared to pay a premium for convenience and services such as music consumption:

> just offering voice is not good enough. Going forward it's going to be much more customer focused. Youth want this, you know, young adults want this. Older adults want this.
>
> (Head of strategy, Orange)

Orange spent £4.4 billion on its 3G licence in 2000, and the head of strategy admits:

> we were ahead of ourselves [...] there was a level of naivety there, which looking back on it, we'd spent too much money. We didn't structure the deals correctly. We had no understanding of what... no real understanding of what the customers were going to use it for.

Many years later, the music and mobile phone industries are still struggling to find solutions for music on phones:

> It [*music*] is in the package, it's a key component, but you're not going charge by track or it's not going to be split out as a line item in the tariff, it's just there, you just get it, which is what you're going need if you really want to penetrate the mobile market. Then that sort of conversation I think needs to take place, because you're actually giving your crown jewels over to another business and then you're relying heavily on that other business to maximize it.
>
> (Head of strategy, Orange)

You go and talk to Vodafone and their opening line is, 'you've got to bear in mind that we're the 800 pound gorilla in this relationship'. So when everyone tells you that...you know, I never liked bullies since I was at school, so tell them to fuck off and then you know...So the reality is that they are basically a telephone company. You know, they're nothing other than a distribution platform, nothing other than that, and if they want to be something different then they're going to have to come up with a different strategy.

(EVP/CFO Universal Music)

The record majors take a pessimistic view of mobile phone companies' ability to embrace content:

the Oranges of this world, they're just pipes. And you should not see them as more than that. [...] they have a cultural block. They're not going to be able to do it. You'll have to provide the content to them. And hopefully the technology is moving quickly enough that you know, you'll have broadband and mobile. And they'll go back, even though they've never really come out of that, to what they are, a pipe. They will supply. And for as long as they get their bits of information and get paid on there they will. I do not expect a proactive strategy from out of the mobile operators. What is more intriguing is the Nokia approach...trying to provide dedicated service and exclusive services to their handsets. Will that work? There also you have a huge cultural divide.

(Former CEO, EMI)

Everyone has got a view about what we should and we shouldn't be doing...you've got everyone from Nokia to Vodafone to every Tom, Dick and Harry all thinking now that they're music experts and they ought to be in the music industry, which is wrong too because they can't cut it. They just can't.

(EVP/CFO Universal Music)

The collaborative failure between music content providers and phone companies may have something to do with the fact that both industries have products valued by customers and businesses built on exclusive control of distribution, as the following exchange demonstrates:

Researcher (JW): You can control your world far more than any other sort of media operator.

Head of strategy, Orange: Oh God. We talk about it as if it's the same thing. It's not. The mobile phone network is a very proprietary network, far more so than the fixed [*line internet*].

Researcher (JW): It's always struck me as ironic that, given that you've got two industries which actually do respect control... having control over distribution, that it has been so difficult to talk to each other in a sense.

Head of strategy, Orange: That's a very good point actually. I don't think the conversations we've had have been at the right level. I think now is a... probably the right time... yeah.

There were some grounds for hope that the relationships could improve. With the growth in 3G mobile phone subscriptions, the number and the sophistication of music offerings on phones increased. Nokia's 'Comes with Music', whereby a Nokia handset comes with free access to a digital library, raised high hopes in 2008:

> our bet is that these handsets, that these devices get replaced annually and on the basis of a handset getting replaced annually, every time a handset gets replaced we will get paid. [...]. But you know what, it's an experiment. They're [*Nokia*] doing it in a big way with a large number of handsets, in certain countries to begin with, with a massive marketing campaign. You've got to bet with them. It's worth the bet. It's worth the experiment.
>
> (EVP/CFO Universal Music)

This is just one example of the dynamic, and possibly confusing, media and communication choices with which the consumer has been faced. As it turned out, the consumer was confused and Nokia's *Comes With Music* service closed in 2011, the same year as Nokia's announcement of its, with hindsight devastating, partnership with Microsoft. The retail and wholesale pricing relationships between handset, call-tariff and content continue to trigger complex negotiations between handset manufacturers, mobile operators, and the music industry as they fight for the power to define value and as of 2013, no phone company has managed to establish a compelling music service with a notable market share.

Having now outlined all the areas of shifting values which were identified by the research participants, the next chapter pulls together their alternative and competing themes of cultural and economic protection.

4
Custodial Tensions

In using the word custody to group diagnostic themes, I am drawing from its definitions of *safe-keeping, protection, care* and *guardianship*. A key question raised in this chapter is 'against what or whom is protection being sought?' Later chapters will reflect more critically on that question, but in this chapter I present edited extracts of the original research texts to illustrate the various claims for protection. The custody theme is characterized by polarities, tensions and dilemmas, which are summarized in Figure 4.1:

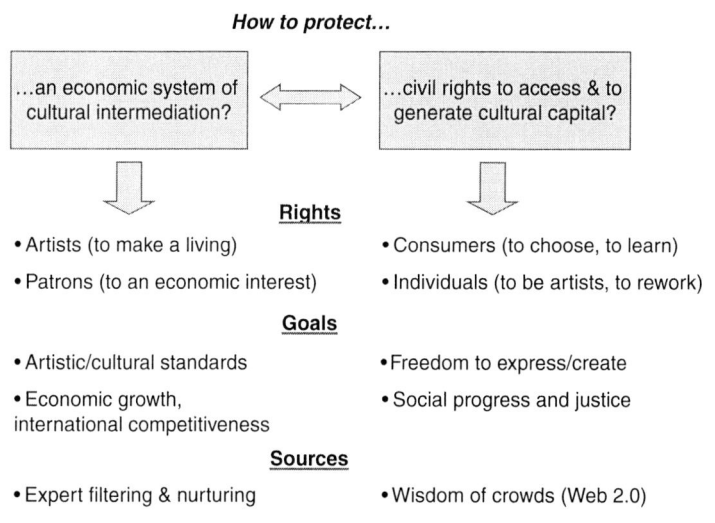

How to protect...

| ...an economic system of cultural intermediation? | ⟷ | ...civil rights to access & to generate cultural capital? |

Rights

- Artists (to make a living)
- Patrons (to an economic interest)

- Consumers (to choose, to learn)
- Individuals (to be artists, to rework)

Goals

- Artistic/cultural standards
- Economic growth, international competitiveness

- Freedom to express/create
- Social progress and justice

Sources

- Expert filtering & nurturing

- Wisdom of crowds (Web 2.0)

Figure 4.1   Custodial tensions

In illustrating these tensions and dilemmas, there are two connected questions which draw out apparently conflicting views of culture and human nature. The first one refers to *cultural custody*, such as the patronage of artists, and the guardianship of the cultural legitimacy of art, fashion and taste, versus the protection of individual freedoms to access and enjoy cultural products, and to build on them in a derivative way to create, innovate and express in ever new ways. The central question here is: 'who should influence the music choices which we can enjoy?'

The second element refers to *economic custody*, such as the protection of recordings, files and intellectual property assets, and the control of distribution channels. Here, the central question is: 'to what extent can we be trusted not to take something without paying for it?' or 'how can you get people to pay for something which appears to be available for free?'

As one might expect with dilemmas which have been the source of so much unresolved disagreement, some research participants maintain a consistency of argument on one side or other of these questions, whilst others jump from one side to another. What follows are some textual examples of the various arguments. In order to give prominence to the text, they are presented at this stage with minimal commentary and critical analysis.

Cultural custody

So why EMI is fucked is not only because of their snobbish view on records. You know they wouldn't sign Leona Lewis who's gone on to sell the biggest selling debut of all time because she's a TV reality star and (EMI executive) doesn't want TV reality stars [...]. The (the music market) is no longer now decided by 3 or 4 blokes in a meeting carving out between them.[...].. those record companies are run by, and this sounds like my old mantra, white, bald, probably protestant, possibly labour voting, heterosexual men and you know, you cannot have an industry that doesn't appeal to...[...] Everyone that buys the product are not those people, by and large, you know, so it's dominated at almost every level by those people. That's when I knew that it was not going to succeed going forward and you're right...democratization, you're calling it...how can it be with those people at the top? EMI just signing middle class boys that went to public school playing guitars. I mean, who cares? You know. But they do because it's clearly for them, you know.

(Independent producer/A&R)

This quote colourfully captures a view that the traditional role of record companies as cultural intermediaries is becoming obsolete. The participant argues that the major record companies have lost any cultural legitimacy they may (if ever) have had, through complacency and lack of diversity in their management. A possible response to this situation, illustrated by the corporate strategy manager at Nokia, is the promise of plurality from technology-enabled diversity:

> I don't really trust in any one person or company to do the job of the thousands, and I think...nobody is as smart as everybody. [...] I would rather have...an identity here which I can plug in if you like into an open system where it can be, 'oh, I've discovered that I actually do like things that have elements of bluegrass and a bit of this and a bit of that a bit of the other'. And then just throw it out into the web 2.0 world and see what sticks.
>
> [...]
>
> I think there's a huge market for sort of 'curator commerce', people who are looking for help in the [...] there's so much choice and so many confusing decisions for individuals that they're looking for a trusted guide, a curator. [...] It's much harder to have the recommendations as a sort of single bottleneck point in the industry which it has been in the past.
>
> [...]
>
> And it might be that one of the metrics that you then have as a parameter to search on is what's hot, i.e. what do other people like, and so that's just a function of how you set up your algorithm for delivering the results. And it might be: 'I don't care about anything except what's in the top 40'. Then that's fine. But if you say, 'that's fine, but hold on, I went to Portugal last year so give me a tweak of that too', that's what I see as the curator commerce point, which is, I am a concierge, becoming much more under my control rather than your control. But the results are still going to be trusted quality music that I like and is arguably of benefit to society.

The desirable world he describes indicates a number of views and assumptions. He acknowledges that consumers need help with their burden of choice, and that such help should come from a trusted guide. Despite his use of human role-based metaphors of *concierge* and *curator*, his preference is for technological over human expertise. He doesn't trust existing cultural intermediaries (record companies), which have been a 'bottleneck', constraining the free flow of repertoire from artist to consumer. His suggested solution combines collective consumer experience and behaviour (the job of thousands) with technology (algorithmic search using preference data supplied by the consumer) to provide an automated service for music recommendation which is under the consumer's control. He assumes that music will continue to be produced and made accessible in ways that these sophisticated algorithmic processes can detect.

This participant's phrase 'nobody is as smart as everybody' is reminiscent of the concept of the 'wisdom of crowds' (Surowiecki 2005). His discursive repertoire draws from what I will call a 'web 2.0' discourse. This refers to those communal and ecological features of the internet which 'enhance, creativity, information sharing, and, most notably, collaboration among users' (Wikipedia). Wikipedia is itself the epitome of the web 2.0 discourse. The web 2.0 discourse features heavily in this conversation, evidenced by the participant's description of Nokia's music strategy as: *helping people discover new music and getting that music to them and helping them connect with their friends and share experiences.* He acknowledges that Nokia does not have the musical expertise to do this, but believes that technology, combined with user experience, can create a more pluralistic offering to consumers than the traditional A&R process, and that society will benefit as a result.

The Nokia manager was one of the younger participants and it was perhaps unsurprising to note that the web 2.0 discourse was also prevalent in the conversation with another of the younger participants, the SVP of marketing at MySpace.com:

> over the last 10 years, like, the whole business of consumerism has gone much more into the hands of the consumers … instead of, like, mass media being able to like pelt you with, you know, an idea, like, a consumer has so much more control.

> I mean, I think that there's always like two kinds of artists. There's artists that like are kind of manufactured […] and then there's artists who are more talented and who are artists […]. I think labels still are really great at delivering that kind of manufactured stuff. I think that the benefit of MySpace has really been for the, you know, more alternative, the quirky … you know MySpace definitely, especially with you know, radio just being more and more kind of homogenous and MTV not really being a music channel anymore. You know MySpace really became the only place that so many … there's huge audiences for bands that some people have never heard of.

MySpace, she claims, has a similar pluralizing social benefit, countering the homogenizing effects of record companies and radio. It empowers the consumer to choose what to listen to, rather than being 'pelted'. It also levels the playing field by giving equal access to all users, thus encouraging 'alternative' and 'quirky' artists who would otherwise find it impossible to get an audience.

By contrast, a third young participant, who was also well versed in the discourse of web 2.0, takes a different line on the wisdom of crowds:

> I just see too much content, and too much content without accurate surfacing for users is bad, and is noise, and you need to get away from noise. And actually the self-fulfilling prophecy is that to get away from noise you go to recommendation. As soon as you get to recommendation you need trusted sources. And trusted sources aren't your peers. The world doesn't rely on its peers. The world relies on experts and as soon as you have experts recommending then it's not very long before experts think, well, I'll take a cut of the revenue and suddenly you start to recreate, and that's not necessarily a bad thing. So a good scenario is that the agents that surface quality content participate in the exploitation of that quality content.
>
> (Head of digital, major record company)

This participant had recently joined a major record company as head of digital, having previously worked as European Director of Entertainment at an internet service provider. Whether or not this view was held at his previous employer is difficult, if not impossible to say, though I did sense that he was experimenting with rhetorical statements to test their impact. A 'good scenario', i.e. that expert agents who are able to 'surface quality content' should participate in its exploitation, clearly sits well within traditional record company sense-making and is a role which most participants agreed was the continuing domain of record companies:

> Our core expertise is finding talent and bringing it into the marketplace. That's what we do.
>
> (EVP/CFO Universal Music)

> I think they (*music consumers*) are going to look for trusted intermediaries to identify and recommend and promote to them what they should be listening to ... that service to me is a very enduring one.
>
> (Former HMV CEO)

I think they (*record companies*) can continue to provide that service because what I wouldn't like to see is that there is no sort of filtered offering to us, because it will just do my head in. I mean, how on earth do you find music that you really enjoy and like?

(New media financier)

I think the internet is an amazing tool but it's still quite unwieldy and I think with musical culture you still have to have your gatekeepers and you still have to have people pointing you towards things.

(Artist manager)

There are many comments along these lines, with record companies constructed as *gatekeepers* and *trusted intermediaries* who are *filtering* and *sifting* and helping consumers *navigate* their way across an ocean of content. These roles are in opposition to the technology-enabled cultural democratization of web 2.0 in that they imply that most consumers are not capable of dealing with choice without the intervention of a cultural intermediary. This might imply a lack of confidence in technology or simply be a protectionist position to perpetuate existing power relations. Some participants were more explicit in this regard:

you say to most consumers, 'there's the biggest toy set in the world, go and enjoy', they wouldn't know where to go. Walk down the street and ask anybody for 10 sound recordings from 5 decades and they can't do it. It's 50 tracks. No chance at all.

(Music service provider)

it was a good thing when there was less choice. [...] I think good for the content and you could argue in a dangerous sort of way, probably good for the punter too. When you give them less choice the choice is going to be better. The content is going to be better. But you're then in danger, and you talk about the philosophical angle of this... there's a danger of the social implications of that

(CEO, commercial digital radio)

there is something of an illusion about choice being great for consumers [...] there is something seemingly compelling about people being able to choose from all the music that ever existed and that has ever been created. [...] I think consumers would simply freeze confronted with hat choice

(Former CEO, HMV)

In contrast to the quote at the opening of this section, these arguments are confidently rooted in the conservative belief that the traditional infrastructure of human intermediaries is still necessary to ensure the maintenance of 'quality content' and cultural standards, that is to say, if left to their own devices, people might not choose 'correctly'. Even where it is acknowledged that technology may replace much of the human filtering process, the process of nurturing talent should not be underestimated:

> If the record labels weren't there anymore then how would...we'd have to find another way of sifting the vast quantities of music that are out there. That's not impossible to do at all and in fact with technology it's much easier to do. So it could be done. But even those other ways of sifting music are still dependent on, you know, record labels finding, nurturing, developing, you know, talent and encouraging talent
>
> (Head of interactive music, BBC)

> Well, I think artist development is still a massive talent of labels
>
> (SVP Marketing, MySpace)

Several participants stress this role of artistic patronage, which they claim is complex and largely misunderstood:

> the record companies have got the difficulty that they've got to manage those egos and actually you need to have...you need to counter an ego with an ego. You almost...you have...you know, just to manage those artists, because they are a bloody nightmare
>
> (Music service provider)

> artists...have got that incredible sense of knowing whether somebody is a suit, or somebody they want to work with. And I've seen it time and time again. And I've seen it with some of my executives who artists wouldn't come near even though they were telling them the right thing. They have that instinct. The second mistake he (Guy Hands, private equity purchaser of EMI) is making is that he is falling in love with the business he bought, right, and that we've seen time and time again. When CEOs start to go to rock concerts and hang out backstage you're in deep trouble.
>
> (Former CEO, EMI)

> you've got to understand certain individuals (*artists*), absolutely vital, and the way they work which may be incredibly idiosyncratic.[...]

You've got, you know, you've got your sausage factory and clearly Eric Nicoli (EMI Chairman) has his biscuit factory, you know, and people get so seduced and just come in and think, 'oh, I can just go a bit rock and... meet a few artists, and it'll just be like that', and they just get it all so incredibly wrong because they don't understand, and so they drive the business into the ground because they try and apply the thing they think will work and forget the things that actually do work [...] The artists can just say, 'fuck you'. You know, because you can't make anyone make a record.

(Independent producer/A&R)

These comments construct the music business as being different from other businesses and warn of the dangers which will ensue if the challenges of managing the creative process are not recognized and handled by people with special skills and experience. The most common outcome of this lack of nurturing is that the talent pool will dry up:

these TV shows I watch now like the X Factor and all that, which you know, everyone's obviously making fortunes out of, but it...to me it's a little bit like what they're doing to the seas, you know, it's like trawling. And a lot of these kids might have a little spark of talent, because they're not having the time to really just play like children should and maybe get really good at it, it's all happening too early and of course by the time they're...once you've been on TV once and your face has been on there, you'd be hard pushed for anything else to happen to your career as far as having a career after that. [...] They're actually not getting the chance to develop and that's going to make the music business suffer because it's just like over-fishing. It's the same thing.

(Composer/producer)

the negative aspect of technology for artists, is that you have this sort of greater consumer demand for newness because people can just access so much more. And artists, it seems, have less time to, you know...there's more pressure to sort of develop faster, you know, as an artist developing and all of a sudden you have these, you know, these younger artists who maybe would have spent 3 years sort of honing whatever it is that they do and they're forced out into a sort of a bigger stage

(MySpace, SVP Marketing)

This premature over-exposure of artists in a world with a voracious appetite for newness may mean the end of long-term 'career' artists, harming not only the recorded music industry, but also the related industries such as live music:

> So, if in the future the recording industry isn't going to be able to develop talent in the way it has done in the past, and you know from your experience the huge cost, then it's going to be a problem for the live industry. [...] There's this great scarcity value of The Police because they haven't played for 20 years, or Led Zeppelin. But we are not going to...we're not going to have that luxury in the future because there aren't long term artists any more.
>
> (IFPI Chairman)

This last quote refers to the long-term impact of lower record company investment in artists due to lower profits resulting from piracy and file-sharing. Another participant sees the consequences as more harmful to the diversity of repertoire:

> moving forward a worry I would have is that as labels get less...you know get more aggressive for less product, because they seem to be signing less, developing less, I would worry that things would get much more homogenized, and I also think that the less product thing is a big issue. I think much less is being developed because people are putting resources into a smaller number of things that they want to have more success [...] I think pricing and distribution is no longer retained by the people that produce artists' records or at least distribute artists' records, and the by-product of that I think, as a fan of probably slightly more left field music, is I think that then dictates what people listen to. Because I think if you look over the last 5 years at the phenomenon of Friday night TV and then Saturday pricing in Tesco and Woolworths, it does drive a certain sort of music to the media and then that gets reflected in what people stack on the shelves in supermarkets because they're looking at their target audience and seeing them as people that might be slightly less discerning and I think it allows...in an era where there's probably a wider variety of music available, more readily available than ever, [...] you would argue that a lot of the higher echelons have been much more conservative.
>
> (Artist manager)

There is an interesting ironic distinction here. Despite there being *a wider variety of music, more readily available than ever*, the participant's concern is that the media, supermarkets, and the *higher echelons* have more of a homogenizing effect on a vulnerable, *less discerning,* audience. It suggests that the participant believes that greater availability and accessibility of music does not have pluralizing effects if those that produce and distribute artists' records no longer retain control of their marketing and distribution channels.

The theme of control of distribution is as much an economic issue as a cultural one and leads into the second part of this section on custody.

Economic custody

Under this heading there are two distinct concepts. The first relates to the question of who controls pricing and bundling-unbundling, the second relates to usage restrictions.

Pricing and bundling

The previous section ended with the artist manager's concern with the influence that supermarkets have had on music consumption, which was largely achieved through heavy discounting of CDs, in some cases to be 'loss-leaders'. The corresponding decline of traditional specialist music retail channels meant that the record industry lost some of its traditional control and power over physical marketing and distribution channels. Similar concerns exist with regard to digital distribution channels. Of all the things which concern the music industry about Apple iTunes' arrival and rapid dominance of the digital download market, two stand out. The first is that, by allowing all album tracks to be sold individually (*unbundling* and *cherry-picking*), the album format was at risk. The second was that, by pricing all music equally in the early years of iTunes, a key marketing tool was being given up:

> The thing that we didn't get unfortunately is...that the majors and the big indies didn't target as I said, the means of distribution, for example iTunes, and the whole fact that it's come down to single songs and the fact that that's dictated by someone other than the record companies.[...] One thing the record companies did from the 50s through to until very recently is they dictated price with retailers. That power seems to have gone now.
>
> (Artist manager)

What would I have liked to have known? I suppose...I'd have liked to have foreseen the timing and importance of the emergence of iTunes, and had an opportunity to think through fully the relationship therefore the record companies had with iTunes, ahead of iTunes deciding for us all.

(New media financier)

The protection of the album format is both cultural as well as economic, as illustrated by the following two extracts:

I'd love to see albums coming back now. Light and shade, dynamics, the whole thing.

(Composer/producer)

One of the biggest problems if you're running a major corporation at the moment is that we face unbundling in the online world. I don't think we can complain about unbundling. That's the laws of supply and demand, that's the effect of the market and that's a proper sound use of technology. The fact that people choose to buy it on a track basis rather than an album basis is very damaging commercially but is not something that I believe we're entitled to complain about.

(IFPI Chairman)

Whilst the unbundling argument has largely been conceded by the industry as being in the interest of the consumer, the pricing issue has been fought much more bitterly. Unlike the unbundling argument, flat pricing goes against economic fundamentals by implying that demand for all music is equal and constant over time. Apple's counter-argument is that complex variable pricing would be an obstacle to consumer adoption of new technology, as the success of the iPod has largely been attributed to the ease of use, in particular the simplicity of the iTunes experience. In any event, price protection is a battle which is difficult to fight in the public domain due to anti-trust sensitivities with the competition authorities around the influencing of consumer pricing. With the exception of the artist manager, participants were thus mostly quiet on the subject. By contrast, they were loud on the subject of usage restrictions.

Usage restrictions

There is a difference between DRM [*digital rights management*] that stops you and DRM that helps you do more stuff. And I think absolutely, if I could trust the record companies to do cool things and give me cool services then I'd be delighted that they knew who I was and knew what my fingerprints were and all the rest of it, but if I think that they're just going to sue me, then I'm not going to want to give them anything

(Corporate strategy manager, Nokia)

For us anyway, it was the protection point of view, of protecting music and the copying and people wanting...greed, you know, human beings. You know, human beings are greedy. 'I want it and I don't want to pay for it'. That's sort of...a lot of kids are like that anyway now. 'You've got what I want and I haven't got it. You give it to me'

(Composer/producer)

What I think our biggest mistake was, and it's true of a lot of industries, was not to look at the consumer behaviour, and not pick up that people were not only file sharing because it was free, but they were also file sharing because they didn't want any more physical product, and they wanted a lot more flexibility with the music available. So if it had been only the first aspect, the file sharing aspect, probably by lobbying and by changing the legal environment we would have finally succeeded or contained it. By ignoring the consumer, that was not containable, because the consumer's always right. And that, you know, I think is a case study in itself, but we reacted according to the old thinking in that we didn't really try to foresee what the consumer wanted and the process got lost. For example, we lived for 4 or 5 years defending like crazy DRM. Well, I'm not sure that the consumer wants DRM, so if they don't want DRM, if you give them DRM product they're not going to buy it. And then they're going to get the same product non-DRM on a pirate site. So we're forcing them to be pirates.

(Former CEO, EMI)

The question of usage restrictions is the most emotive topic in the industry, probably because the discourse which surrounds it is often highly manipulated. Discussion centres on the question of ownership rights, and DRM in particular. DRM generally refers to software which protects digital files and prescribes their rules of usage. Though it could be described as an economic enabler, it is more frequently referred to in the media as an unpopular obstacle to unrestricted usage. The economic and moral dilemma at the heart of the DRM debate may be expressed as follows:

(1) If you *allow* people to do whatever they want to do, they may avoid paying you money.

(2) If, through authorized channels, you *prevent* people from doing what they want to do, they will do it through *un*authorized channels, where available, and may avoid paying you money.

Or, in the words of a participant:

> I changed my mind with a considerable range of thought. You know, it wasn't like, 'yeah we have to do it'. It was, 'if I'm a consumer I will hate that fucking DRM thing'. But on the other hand it was a huge risk. It was what I was calling 'Tesco without a cashier' because that's really what you're doing. And 'please leave the amount of what you bought in the box at the entry and thank you very much'. So we had a hell of an internal debate on it.
>
> (Former CEO, EMI)

I refer of course to the taking and the giving of free music. It is not a new phenomenon: copying and sharing of LPs, tapes and CDs have been common domestic practices for decades. It threatened the recorded music business but, until 1999, did not prevent it from growing. A plausible story compounded by public discourse proposes that, in the last decade, developments in technology and media have made fundamental improvements to (a) the quality of copies and (b) the ease with which copies can be made and shared. Consequently, as research indicates,[1] two-thirds of recorded music consumption in 2007 was via unauthorized channels. There are problems with this statistic, though here I'm concerned not with its reliability, but with the way in which participants describe authorized and unauthorized usage.

A participant recalls that record companies began their policy-making journey with an assumption that the environment was controllable:

> it's a mental thing and I remember I think it was Rupert Perry (*EMI CEO*) in one of those early meetings. He said, 'but we have to own distribution'. I said, 'Rupert, in the internet world you cannot own distribution'. He said, 'why can't we?' I said, 'because you can't afford it'. He said, 'what do you mean?' I said 'the global music industry: £35 billion, IBM's turnover: £175 billion; Microsoft, Oracle, AT&T...,' I said, 'you're tiny. With the greatest respect, it's tiny. And these people, even they don't own distribution, right. It's called the internet, so think about the people you can partner with'.
>
> (Music service provider)

For many years the record companies pursued solutions which would, if not control the internet, then at least protect their content from unauthorized usage. Part of their frustration pre-dates the internet and is rooted in the fact that the unprotected CD format was developed more than 25 years ago, before anyone could foresee that PC-based ripping and burning capability and internet file-sharing would be common domestic practices. Despite the fact that the CD is an open format allowing users to do whatever is technologically possible with the digital music they upload to their computers,[2] the industry has ironically attracted much negative sentiment and criticism for its perceived desire to control the usage of its content. This fact still causes frustration and confusion for many in the industry, especially when they compare the situation with other audiovisual technologies which were developed later, such as the popular DVD format for movies, which are universally copy-protected:

> it was worth a try, and it is still not clear today why every other type of product sold on the equivalent of a compact disc is DRM-protected, and the only thing that is not is music
>
> (New media financier)

> Rupert Murdoch should probably have been running the music business because I notice now that I've got the Sky thing, the recorder, that I can record it, but no way can you copy it. I can video a whole show and people say, have you got the video of this concert? They have to come round to the house and watch it because I can't do them a copy. So he's managed to do it. Why couldn't the music business manage to do it? I don't understand
>
> (Composer/producer)

> There was nothing wrong with copy control protection, other than one huge problem: it didn't work.... it's still amazing to me that we've put a man on the moon, we've done all sorts of extraordinary things and from the days of the cassettes all the way through to the days of the CD we haven't been able to find some copy control mechanism which protected the CD ultimately.
>
> (IFPI Chairman)

The argument is one of practice rather than principle. Anything which degrades the prevailing consumer experience or restricts something they have become accustomed to enjoying is going to be problematic, however consistent it may be with other technologies and with the law. It began as a dilemma regarding the protection of a physical product (the CD), and migrated to the domain of digital distribution, once legitimate digital download services began to generate meaningful revenues. Though it was by no means the first service, iTunes became the world's largest music retailer and attention surrounding DRM became focused on Apple and Steve Jobs. The DRM issue is often confused with the question of interoperability, which in this context is the term used to describe the ability to play music acquired from any source on any device. It became contentious because Apple, Microsoft and Sony all developed digital music players with proprietary software which restricted usage of songs bought via certain services to certain players. The popularly quoted reason for making consumption of music 'device-dependent' was that the margins from device sales have been considerably greater than the retail margin on music sales.

The practice has drawn criticism for being anti-competitive, for example from the French government, who proposed legislation to force Apple to make their FairPlay software code available to other suppliers. In an open letter posted on the Apple website in 2007, Steve Jobs made the argument that he was bound by contractual obligations to the record companies to protect their music. Deflecting attention away from FairPlay, he invited record companies to 'abolish DRMs entirely ... this is clearly the best alternative for consumers, and Apple would embrace it in a heartbeat'. In supporting this proposal, he made two key points: firstly, that record companies already sell over 90% of their music unprotected (via CDs), and secondly that 'DRMs haven't worked, and may never work, to halt music piracy' (Jobs 2007). The letter attracted an enormous amount of media attention, some anti-Apple, some anti-music industry, and most pro-consumer. EMI, who until this point had been one of the most vocal proponents of copyright protection and DRM solutions, reversed its position by announcing, less than two months after Jobs' letter, that in partnership with Apple, it would start selling digital downloads free of DRM restrictions 'addressing the lack of interoperability which is frustrating for many music fans' (Nicoli 2007).

Though I had been an internal advocate for dropping DRM, my interpretation of EMI's motives are that this was less to do with doing the 'right' thing for consumers, and more to do with striking a tactical deal. EMI was unofficially 'for sale' and desperately wanted to generate positive press coverage and to be perceived as a strategic first-mover. Apple needed to make a concession to divert attention from the French government's attempt to force interoperability on the 'closed' technology of the iPod. This EMI-Apple announcement was a mutually beneficial but isolated deal in an otherwise difficult relationship between Apple and the music industry, and EMI's move was unpopular with the other major record companies:

> The problem to me wasn't DRM, it was Steve Jobs and inter-operability. Now I don't think many people believe me when I say this, but I do strongly believe it. It had got to the stage where Steve Jobs believed it was untenable to keep going and his letter was ... I think the last flailing of a dying man or a man who was going to give up on this [*prior refusal to adopt an interoperable DRM*].
>
> (Chairman, IFPI)

> DRM and the issue about interoperability and is it the record indus-try's fault that there isn't interoperability between Apple technology and Microsoft technology and my view is I feel incredibly flattered that people think we're so important, [*laughs*] that it's actually down to us that there's no interoperability between Apple and Microsoft. It's just ridiculous, you know, it's completely ludicrous, but Steve has spun a good story. He's spun a phenomenal story.
>
> (EVP/CFO Universal Music)

The debate goes on and many record companies continue to insist on DRM for sales of digital downloads, eliminating any real benefit for con-sumers as there is no clear indication on the iTunes store as to whether the song you are buying is restricted or not. An indication of the difficult relationship between Universal and Apple is that Universal, for a time, continued to sell DRM-restricted on iTunes whilst allowing DRM-free downloads on competing download sites.

Apple's commitment to interoperability is called into question by another participant with reference to the iPhone:

> You're really messing with the consumer because you're tying him ... I mean if I was anybody with normal earnings I wouldn't go near the thing (*the iPhone*). You're tied to 18 months contract and you're tied to buy a very expensive machine. I think it's far too ... and you're tied on top to the iTunes DRM. So you can't really transfer your music into anything else. I think it's one step too far, because where I see the iPod was clearly a winner in terms of design and technologi-cal advance and all of that, now the world has changed in 4 or 5 years. And I think people will hate it. And I wouldn't be surprised if the iPhone was a failure. From a technology point of view it's great but there are a lot of people on its heels, you know, Nokia being one of them. And I strongly believe, and I think we've seen it in music, that closed systems are doomed. Now that has been the Apple philosophy

forever and really the computer, the Apple computer, really started to take off when it accepted Windows.

(Former CEO, EMI)

With hindsight, this view may have been wrong about the competition from Nokia, but the jury is still out on whether Apple's 'closed system' will prevail in the long term against more open systems such as Google's Android platform.

The participants were evenly split on the question as to whether DRM solution providers have a positive economic future. Those who still believed in it did not have the answers, but were relying on the moral or social obligation of intellectual property protection being a sufficient driver of innovation:

> DRM solution providers...I think that's the way forward. You've got...at some point you've got to work out how that all works properly.
>
> (Independent producer/A&R)

> the fact that people are allowed to basically pirate at will today...something needs to be done about that because it's kind of wrong. It's not <u>kind of</u> wrong. It <u>is</u> wrong. [...] We've got to try and protect [the music] until something else comes along
>
> (EVP/CFO Universal Music)

The alternative view is expressed in richer metaphors:

> if you take locks and keys off the content then it's like letting the foxes into the henhouse. And you think, hold on, are they going to run away with all our...[...]...there's a big old back door which they're kind of occasionally trying to paper up with suing individual customers but it doesn't really help [...]...if it (music) is somehow set free and [you] allow people to innovate and use it and share it and mix it up much better then I think that would provide a lot of options for other innovation
>
> (Corporate strategy manager, Nokia)

> the genie is out of the bottle in terms of pirated content. And there never...you know peer to peer, illegal downloading is never going to stop in my view unless it becomes a) socially unacceptable, difficult or b) the record companies offer something which does the job instead, that makes it just not worth the effort
>
> (CEO, commercial digital radio)

the truth of the matter is, and President Sarkozy in France got this absolutely right...He decided to help the music industry...He said two things. First of all the internet cannot be the Wild West for intellectual property. That's just completely wrong. He then secondly said, it cannot be right that something you purchase legally can be played on less devices than something you purchase illegally. And that's a very perceptive comment from a man who's got many other things on his mind. He says, if somebody goes and steals it they can play it on all these devices and if somebody goes and purchases it legally from iTunes they can only play it on their iPod. He says that can't be right. Now...and he put the fault at the door of DRM. To my mind the fault is at the door of inter-operability. But that particular genie is out of the bottle and we've moved on.

(Chairman, IFPI)

With regard to the French and UK governments, the music industry has 'moved on' to another approach to combating unauthorized usage which introduces another metaphor. This is the 'safe-harbour' or 'mere conduit' provisions protecting internet service providers (ISPs) from any liability arising from the illegality of their users' activities, such as unauthorized file-sharing of copyright content.

as the internet was developing the concept developed of a safe harbour for service providers on the basis they couldn't be responsible for everything that went over their pipes and cables. And that was a reasonable thing to do as the internet was needing to find some oxygen and flourish. But at the end of the day it became an opportunity for people to turn a blind eye to flagrant infringement and that's been very damaging to the creative industries [...] When you then have a situation where they know what's going on, you draw it to their attention and you offer them a solution and they then ignore that and are still able to hide behind a safe harbour, that's not an equitable world. And nobody would have foreseen that or nobody would have been able to justify that. [...]

This connection strategy which Sarkozy has gone for has been one of my overriding principles, at times against my members, because a speech I've given to all the top politicians and media around the world is really as simple as this. In the John Kennedy household there is a 17-year-old girl, a 15-year-old boy and a 9-year-old girl. If the 15-year-old boy was stealing music on the internet and if as a result he got the internet connection disconnected his sisters would kill him and he would be much more afraid of his sisters than he ever will be of the IFPI. And to me that's the dynamic that's going to save the music industry because that would be repeated around the world. He would not... for himself he would not take the risk of losing his internet connection. People just can't anymore. So if you could introduce that dynamic for a very large percentage of people you would stop the stealing of music.

(Chairman, IFPI)

The Sarkozy connection strategy referred to in this extract is known as Hadopi after the acronym of the French government agency created to administer it. It is an experiment pioneered by the French government in 2009 which involves cutting off the internet connection of persistent file-sharers if they do not respond to warning letters. Despite millions of letters being sent out, it took until June 2013 for the first suspension, for 14 days, of an offender's internet connection. At the time of writing, this remains the only such suspension. Measures of the impact of the policy on illegal activity are inconsistent, leading current French president Francois Hollande to consider abandoning Hadopi. Meanwhile, a similar policy has more recently been implemented in the US.[3] The UK Digital Economy Act (2010) provided for the same approach, though it currently seems unlikely ever to be implemented due to conflicting stakeholder lobbying and arguments over who should pay for it. Nevertheless, one participant felt that the French were more culturally inclined to accept such a solution:

This will happen in France. It will happen in France. It absolutely will happen because it's a country that respects all things cultural.

(EVP/CFO Universal Music)

As illustrated at the beginning of this chapter in Figure 4.1, the question of protection for cultural products is one which will always be a dilemma to be reconciled. The implication from this participant's comment is that, in the UK, culture is less of a priority than the principles of liberty and privacy. The participant Lord Triesman, who at the time was the first government minister for intellectual property, indicated that the UK government was watching the French experiment closely and aimed to broker a similar voluntary solution between the music business and the ISPs:

And we've said if there isn't a business model we'll legislate and we will. The plan is to legislate.[...]

There will be some difficulties. Some people are saying, if we go along that route it will mean that clients of ours will become known to you. There are privacy issues. Well, so there are I guess for paedophiles, if you argue that to its conclusion. But no one will accept that conclusion. So the new balance I think, has to be struck between doing things about people who steal property, doing things which provide again the right environment for new business models with streams of income which people will accept are legitimate, and doing things for consumers which mean that they're not impeded by ... well, they're not threatened to be impeded with rules you probably can't apply in any case. So that's how I'm approaching it.

(UK government minister)

Exactly six months after this interview, the UK government announced that it had brokered a 'memorandum of understanding' between ISPs and the music industry which commits both sides to explore solutions, with a view to avoiding government legislation. The announcement made news headlines across the UK media and the following edited transcript from BBC Radio 4's 'Today' programme[4] captures the themes nicely:

- **Evan Davis (presenter):** 'at some point, surely, we're going to have to go in, monitor what people are doing, and say you can't illegally share files'
- **Billy Bragg (left-wing musician):** 'are you really? ... there's nothing better than peer-to-peer recommendation ... we're going down the wrong road if we criminalize our audience ... it won't help musicians make a living'
- **Andy Burnham (government culture secretary):** 'the solutions are out there that will work for everybody. We just have to find them ... We will legislate if we can't broker a solution'
- **Becky Hogge (director of OpenRights Group):** 'government is not giving the music industry enough incentives to give consumers the choices they need'

In the absence of a representative of the recorded music industry, Evan Davis opens with the assumed music industry position. Billy Bragg replies in a way which indicates that record companies' interests are not necessarily aligned with the interests of their artists. Andy Burnham claims the role of referee ensuring that 'everybody's' interests will be respected. Becky Hogge challenges that the solution Andy Burnham is brokering is only between two powerful economic parties (ISPs and record companies) and that consumer rights require more protection from government.

Five years later, the ISPs and the music industry are still arguing and nothing concrete has been implemented. This example, of many parties seeking protection but remaining deadlocked, is a recurring theme of the book. Part III will examine its narrative power, and Part IV will draw out the industrial and political implications of continuing deadlock.

5
Hindsight and Strategic Sense-Making

> one route to market [...] with hindsight, is a ludicrous proposition.
>
> (EVP/CFO Universal Music)

In this third and final section presenting the most common themes and concerns of the participants, I illustrate some of the many participant retrospective references to blinkered views of the world, whether their own or those of other stakeholders.

As mentioned earlier, participants were asked what, with hindsight, they would like to have known ten years ago, inviting them to share any cognitive shifts:

> Well 10 years ago we were just about to buy a 3G licence for about £4.4 billion. I'd have liked to have known what I know now in the auction. Because there's no way [...] I probably wouldn't have...you know [...] I was that guy going out to the Playboy mansion, Hugh Heffner's pad and saying, I want your global adult deal. I was that guy going to the IFPL and saying, I want your sports content for my mobile phone company. And there was a level of naivety there, which looking back on it, we'd spent too much money. We didn't structure the deals correctly. We had no understanding of what...no real understanding of what the customers were going to use it for. [...] A learning from that is that it's taken a lot longer to get people to be aware of what a mobile device can now do, more than just make a phone call. So in short, we underestimated the length of time to change music behaviour, and it's taken 3 or 4 times as long to get people to use more than just SMS but using MMS and starting to share stuff. And we were ahead of ourselves. So we could have [...] not focused on all the sexy stuff, focused on making the basics right,

then introducing the sexy stuff rather than trying to do it altogether. So that's what I, you know, that's what I probably would have done at a high level. But that would probably have got me out of a job. So I was doing all the sexy stuff. [...] so in answer to your question, I wouldn't have pushed as hard for 3G products. I would have focused on the basics. But I would have lasted about 5 minutes on the Board if I'd said that, so you know, it all gets wrapped up in this big hype.

(Head of strategy, Orange)

The participant has elsewhere referred to his frustrations that technology has 'over-promised and under-delivered', and there is no better measurement of that mismatch than in the (with hindsight) wildly excessive (and for some phone companies) crippling amounts which were paid in the UK and European auctions for third generation (3G) mobile spectrum licences in 2000. Note, however, the participant indicates that someone with a prophetic vision of the future would not 'have lasted 5 minutes on the Board', such was the robustness of the mutually constructed 'hype' around the impact of 3G technology on mobile phones.

Commitment

The bidding for a 3G licence is a good example of how, according to Weick (2001), post-decision behaviour, based on choices which are both public and irreversible, produces a powerful behavioural commitment. This commitment marshals forces which destroy the plausibility of alternatives (which may have been considered plausible prior to the decision being made), and removes the ability of those alternatives to inhibit action. The mutually constituting discourse which emerges (e.g. 3G must be the future because so many companies were prepared to bid for it) may cause a 'deviation-amplifying loop' (Maruyama 1963, cited in Weick 2001, p. 297), which has an equal capability for mutual destruction, as was the case with regard to shareholder value in mobile operators. The 3G auction story is also a sobering example of the power of the discourse of technological inevitability which is promoted by corporations (and governments) to justify continuing liberalization and marketization of the communications sector (Murdock and Golding 1999; Hesmondhalgh 2002).

By contrast, the Napster phenomenon was indicative of a different but equally powerful discourse, which was, at least for a while, outside the influence of big corporations and government. Three participants referred to Napster in their responses to the hindsight question. None of them were record industry participants, which is perhaps not surprising given that the prevailing industry view of Napster, as a dangerous band of pirates to be destroyed, was a perception that has not (yet) needed to be changed with hindsight. This is another example of how the commitment to a course of action, in this case an industry-wide rhetoric of the criminalization of file-sharing along with a policy of litigation against offenders, has shut down the possibility of file-sharing as a legitimate activity being plausibly and constructively evaluated.

One participant's view, which seemed to justify the record industry approach, surprised me:

> **SVP Marketing, MySpace:** the big missing piece of 10 years ago was...you know, that was sort of when Napster launched. I don't think that anybody saw taking free music becoming such an accepted practice of a generation of people then. I think people sort of thought, 'oh god, our music's being pirated'. But like, nobody shut the pipes off, you know, or quick enough. You know I don't think anybody...
>
> **JW (researcher):** Well do you think they should have done?
>
> **SVP Marketing, MySpace:** I just don't know if you could have predicted how widespread piracy would have become.

Her view, that shutting off internet access might have been justifiable had it been known how widespread socially acceptable (at least to a younger generation) piracy would become, is a rarely expressed one from someone of her age (under 30 at the time of the interview) and sector.

The more common view is that expressed by the Nokia participant:

> I remember I was in Boston when first Napster came out, and I thought it was fantastic, as did the rest of the world, apart from the music industry who were: 'oh my god!' They just saw it in a negative way, whereas if I had my time again, and I 'got' Napster when it came out...I'd love to ask, and this could be one of your questions to your other music executives, what would you do this time? Would you actually licence it and have an amazing amount of cash and amazing amount of innovation and excitement, people listening to music they've never heard of before.
>
> (Corporate strategy manager, Nokia)

This echoes another participant's view of the missed opportunity

> **Music service provider:** They made billions out of it [Napster]. They made so much money out of the advertising revenue that they got by allowing…
>
> **Researcher (JW):** What, the initial Napster?
>
> **Music service provider:** It was a quarter of a billion. I can't remember where I saw it.

My doubting question to the participant was based on my understanding that very little money was made out of Napster prior to it being forced to shut down,[1] which is why, to my knowledge, there was never any material offer made to the record companies for their collaboration in, or licensing to, Napster as a commercial venture. It seems unlikely that dialogue would have progressed even if there had been cash on the table, given the fierceness of the recording industry's determination to shut Napster down. However, I mention it because there is widespread belief amongst people who adopt a web 2.0 discourse that Napster could have been successfully commercialized. They believe it could have transformed the record industry in a positive way, had record companies not managed to shut it down, and instead had they collaboratively reconceived their product offering. When each of the last two participants quoted were asked how this might have occurred, both responded with legalized file-sharing variations of the subscription or flat-fee licensing model.

These views could be interpreted as implying that Napster's misfortune was entirely related to timing, specifically to the amount of elapsed time for file-sharing to be reconceived as an opportunity rather than a threat. It suggests that inventors who take the time to conceptualize and articulate new products from new technology are more successful than inventors who simply make the technology available as quickly as possible, without being sensitive to the competing interests and value-systems of powerful stakeholders. However, it is questionable to what extent Napster can be credited to an inventor at all. In their paper *Contested Codes: The Social Construction of Napster*, Spitz and Hunter (2005) 'document how popular, or vernacular, theories mobilize around a tool or technology to reveal its "true" qualities' and they 'demonstrate how and why certain (subjective) meanings increasingly take on the status of truth while other (equally subjective) meanings are pushed farther out to the fringes.' They challenge the 'one-time heroic event' depiction of Napster inventor Shawn Fanning, in favour of a view which sees

Napster as a social process 'constructed within a culture already attached to certain values', and as 'an organizing principle for widely diverse issues', illustrating their point with reference to Al Gore's appropriation of the construct to suit his political agenda. Whether featuring as protagonist or antagonist, Napster has been given a powerful symbolic role in 21st-century storytelling.

Historic success and cognitive impairment

One view of strategy (Burgelman 1983) is that it is a theory about the reasons for past and current success of the firm. This resembles De Bono's (1984) definition of strategy as good luck rationalized in hindsight (p. 143) and also recalls Weick's views on retrospective sense-making and the destructive consequences of eliminating plausible alternatives to committed strategic choices. Closely related to these concerns, there were many participant observations that historic success breeds complacency and short-sightedness. Record company personnel are frequently described as mistakenly believing that they are in control of the market, and continuing to work on the assumption that the knowledge, instincts, skills or structures to which they attribute past success are enduring and will continue to sustain good fortune. In some cases, record company participants themselves recognize this through reflection. To the hindsight question one participant responded:

> Probably to know that I don't know, because when I reflect on the last 10 years I think people had a lot of certitudes about the future and pretty quickly the future was changing much faster than we thought. [...] I...we thought that we were working within the framework of the old model and going into an evolution while actually there was a revolution taking place. So we didn't see it. Now I don't think anybody saw it [...]...we reacted according to the old thinking in that we didn't really try to foresee what the consumer wanted, and the process got lost
>
> (Former CEO, EMI)

The view from outside the record companies is similar, though less charitable:

> It's like, kind of the fall of the Roman Empire [...] record companies, they're almost Titanic-like, heading towards an iceberg. They won't

give the market what it wants. They think they can dictate to the market. That is just so sadly incorrect, especially now when those four majors have lost their dominance. They can't dictate anymore in the way they used to be able to, you know. And so, not giving consumers what they want, to me is just like in any other industry, it's insanity.

(Independent A&R/producer)

This participant is the most colourful in his descriptions, elsewhere adding King Canute and the ostrich to his list of analogies for record company behaviour, but long-term success and complacency are regularly cited by others as a diagnosis for perceptual insensitivity:

Music has always been the coolest and the sexiest business to work in, and all these things and people in the music business have never had to sort of, you know, think about being a little bit more 'out of the box', because it's just always worked.

(SVP Marketing, MySpace)

This perception that the industry business model has 'always worked' without being required to innovate or respond to a changing environment introduces a feeling of resentment by others that record companies have had a comfortable and lucrative life for too long, and are to some extent responsible for their own demise. Consider this anecdote from a participant recalling a meeting with a senior executive from a major record company in 1998 (notably, the last time the participant had had any dialogue with any record company):

that might lead to electronic retailing of music, hear it, buy it, burn it, type of thing, but that's what it was called then. And I sort of laid this out as the likely future that I could see. And I never forget, I can't remember who it was but a guy said, 'we will never do electronic retailing of music'. Really? [...] 'We will never do electronic retailing. We will never allow people to download our music'. Why not? 'It would damage the sales of our CDs'. And everybody then tucked in behind him and that was the end of that. And I just thought, well, you know, it'll rain on you. And it did, you know, now, for lots of different reasons. But I thought that kind of ... if that's symptomatic of record company thinking, they deserve everything they get.

(CEO, commercial digital radio)

There is a connected theme that such complacency breeds corpulence. Record companies, and symbolically, their executives, have become overweight relative to the resources which support them. Descriptions of record executives as 'fat' were commonplace within the research conversations, with multiple references to 'fat bastards' and 'fat cats'. Whether referring to excessive size or wealth, there were many instances of record companies being described as too big, with comments such as 'there is no need for a heavy infrastructure', 'every record company needs to reduce its size because it's just not necessary', 'we'd fire half the staff', and 'to get through this [they] need to be a lot leaner'. There is also a wider public discourse of injustice at the salaries and extravagances of record companies relative to the poverty of most artists, invoking the argument (e.g. Miege 1989) that the permanent oversupply of creative workers means they bear the unrecognized costs of creation and production by being willing to forego the benefits of secure working conditions. For all these reasons, the perceptions of excess, resentment, injustice and ridicule[2] continue, as evidenced by the frenzy of media coverage following the acquisition of EMI by private equity company Terra Firma in 2007.

In understanding the origins of these fat-reduction recommendations, it is useful to consider the context of recent history. The last cycle of economic success in the recording industry is generally attributed to the rapid growth of the CD format in the decade leading up to the recording industry's global revenue peak in 1999. In addition to stimulating sales of new products, record companies also benefited from consumers buying again on CD products they already owned on vinyl or tape format. This led to a popular view that the record companies enjoyed excessive profitability, and an over-reliance on one narrow business model:

> it took the rapid decline of the CD to focus our mind [...] We made 'too much' money, let's say [...] The big buck was in the CD [...] That was the big money so you chase after that and I think [we] wrongly excluded everything else.
>
> (EVP/CFO Universal Music)

This participant acknowledges that the great success of CDs had consequences for cognitive faculties, impairing strategic vision. This could be interpreted as confirming the theory that complacency, stability or equilibrium are precursors to demise, and that survival or change only happens on 'the edge of chaos' (Pascale et al. 2000). Expressed another way, living things face constant threat, and must constantly seek higher levels of mutation, experimentation, from which new solutions are born. An alternative, but similar sentiment is the popular change management metaphor of the burning platform which is required to mobilize action. Either way, it took 'stark and desperate' circumstances for the industry to realize 'the reality': that a business based on 'one route to market', i.e. ignoring all the other ways to derive economic value from music, 'is with hindsight a ludicrous proposition'.

Believing your own rhetoric

Organizational leaders face a dilemma. On the one hand, they are expected to project sufficient stability and continuity of vision to inspire others to follow. On the other hand, the danger of excessive commitment to one strategic choice suppresses the possibilities of experimentation and the exploration and balanced discussion of alternative courses of action. This is well illustrated by the response of the HMV CEO to the question of what he would have liked to have known ten years earlier:

> I think one of the dangers when you're the Chief Executive, particularly of a public company, where you so frequently have to stand on a stage and persuade people that your strategies are right, is that you can almost end up believing your own rhetoric. And there were a number of things which we would consistently articulate to the investors which I think with the benefit of hindsight were, to a varying degree, fallacies. So, one was that internet distribution as a channel to market would only ever become a sizeable <u>minority</u> part of the market. [...] As you would expect in any sizeable organization you've got a relatively intelligent group of people with differing views and perspectives so there were some people who always believed that the future would be... the future of <u>retailing</u> would be internet distribution, let alone the future of music distribution. There were also some very very dyed-in-the-wool sceptics. I mean one of the problems in a company like HMV is that the senior management have all got there because they're shop keepers so even though one tries to be rigorous about this and open-minded it is very seductive to believe that everyone still wants to use shops and that that business model, as it were, will ultimately prevail. Not only was the senior management shop keepers of several decades duration, a lot of the senior management have worked in the music industry for several decades and were very wedded to the story that what consumers wanted to

do, and would always want to do was to build collections, and you know, the collector mentality was very much something which we overlaid on consumption of music.

(Former CEO, HMV)

Leaders are often denied the luxury of vacillation, as this participant points out. Once you've chosen a strategy, you stick to it, at least for a while, in order to lead both your employees and your investors with confidence. The danger is that you 'end up believing your own rhetoric' which makes the balancing of reflection, experimentation and commitment to a strategy very difficult. There is a view of such corporate rhetoric, well expressed by Cheney et al. (2004), in which, consistent with theories of autopoiesis (Maturana and Varela 1980), organizations can be seen as organic self-reproducing systems interacting with their surroundings, recreating themselves following their primary goal of survival. Corporate speeches, mission statements, brand campaigns and market strategies are often rhetorical devices in the classical persuasive tradition. As a consequence, organizations continually confirm themselves to internal as well as external audiences. This self-identifying and self-protective behaviour is complex and risky, with organizations often getting caught out by their own persuasive intent, indicating that the strategic impact of rhetoric is not necessarily strictly rational or within the control of the organization. For HMV, whose roots date back to the very origins of the recorded music industry, this had near fatal consequences, with the retail chain going into administration in early 2013.

Having presented the key themes which emerged from the research, in Part III of the book I will go on to examine the language used in a more critical way, highlighting various narratives and identities which can both enable and constrain strategic choice, and the potential for industrial transformation.

Part III
A Storytelling Contest

The next six chapters focus on the language utilized in the discussions of the music industry's past, present and future, with a particular emphasis on the various narratives at play.

For those not familiar with the field of discourse, Chapter 6 provides an overview. It explains the various domains of discourse, and the epistemological claims of discourse analysis. Although this may appear rather abstract and academic at times, I would encourage the reader to bear with me as I believe these principles are important in understanding the significance of subsequent chapters.

Chapter 7 introduces the concept of strategy as storytelling with contextual reference to the music industry. It also considers the role of strategy as a sense-making process for executives.

Chapter 8 uses the original research texts to highlight the most recurrent and relevant discursive objects, specifically: music, the consumer, the record company and technology. These objects are shown to have alternative and sometimes competing linguistic constructions in music industry discourse.

Chapter 9 shows how analysis of industry discourse reveals a series of narratives, each defined by their protagonist. I broadly divide these tales into two groups based on the worldview which they adopt. Tin Pan tales are those which promote a more capitalistic view of the music industry, whereas Wiki tales are products of a web 2.0 cultural ideology. Each participant in the research draws upon more than one of these tales, which is indicative of intelligent stakeholders using narrative to make sense, for themselves and for others, of an environment which is complex, contested and changing.

Chapter 10 takes one of the tales, the Inventor's tale, to explore a fundamental difference in the conception of technology, and contemplates the consequences this has for future strategic possibilities for music now that the industry no longer controls its own technology.

Chapter 11 concludes Part III of the book with a critical reflection on themes of power and ideology at play in the music industry and why winning the storytelling contest is so significant.

6
The Analysis of Discourse

As a finance director, people generally expected me to 'talk numbers'. Indeed, the international language of finance tends to dominate the language spoken within organizations, especially where strategy, performance measurement and management appraisal are concerned. Personally, I have always felt that words are more interesting and important than numbers. Over the course of my career, I found that many organizational problems can be diagnosed by being sensitive to language usage. Given that organizational change initiatives tend to underestimate the linguistic aspects of mobilization or resistance, linguistic diagnosis of particular behaviours can certainly help with the processes of strategy execution and change management.

The domains of discourse

I anticipate that some readers of this book will be unfamiliar with a linguistic approach to organizational studies, so I begin this section with a very brief overview of the field of organizational discourse. As the word *discourse* can cause misunderstandings, it is helpful to start out by identifying categories of discourse that are particularly prevalent in discourse studies. The following domains of discourse are identified in the *Sage Handbook of Organizational Discourse* (Grant et al. 2004):

• Conversation and dialogue
• Rhetoric
• Tropes
• Narratives and stories

Conversation and dialogue refer to contextualized interactions which are produced as part of talk or message exchange. Discourse analysts in this domain are interested in the linguistic and textual exchanges amongst organizational or social actors, with a focus on the *interaction itself*. With conversation, the emphasis is on the extent to which the interaction provides resources for action and further conversation; in dialogue, the emphasis is more on showing the generative (or de-generative) and transformational properties of the interaction: for example, by seeking to show how new meaning is generated via techniques for mediation and mutual awareness. Discourse analysis at this micro-level tends to be the domain of ethnomethodology and discursive psychology.

Rhetoric is a more established and classically defined domain which refers to the devices by which discourse can be used to achieve particular ends. Studies of organizational rhetoric tend to examine the way in which corporate image and strategy are deliberately managed. Of all the domains, rhetoric could be said to be the most self-conscious and accessible to organizational and industrial study.

Tropes are literary devices which are found within each of the discourse domains, but are sufficiently important to warrant having their own category. Metaphor is the best-known trope and is of particular interest in relation to social or organizational change because of its generative qualities. Nonaka and Takeuchi (1995) describe metaphor as one of the three key characteristics of corporate knowledge creation. By creatively connecting abstract concepts to the concrete and familiar, metaphor can stimulate knowledge transfer and provide new perspectives on fixed and mature ideas where existing communication is blocked by prejudice or lack of imagination. Examples of metaphor in earlier chapters include the slicing and baking of pies, to illustrate the division of existing markets versus the creation of new ones, and the honey trap to explain the role of a marginal or loss-making activity in relation to a larger strategic goal. The intellectual property pirate was certainly a metaphor when first used in the 17th century, though arguably there is a point when a metaphor becomes so taken for granted that a new reality is formed. This is considered further in Part IV.

Most tropes are powerful mechanisms for *paradigm reinforcement*. By contrast, irony is a dissonance trope, making meaning visible by highlighting contradiction and absurdity, and by deliberately contrasting the difference between apparent and intended meaning. It can be a mechanism for *paradigm disruption*, i.e. challenging the prevailing view by identifying and articulating the paradox and tensions in life. Irony covers a variety of linguistic phenomena, from the simple verbal irony of phrases like 'as clear as mud' to situational ironies where an unexpected outcome seems to mock the folly or vanity of human intentions, such as the plot of Rupert Holmes' hit *Escape* (aka the Pina Colada song), to use a musical example. Critics of the traditional players in the cultural industries make heavy use of irony, for example in illustrating the self-destructive consequences of content-protection strategies, including the absurdity of suing your own customers. In Part I, I highlighted the irony of how an industry which owes its very existence to technological innovation failed to identify a new generation of technology as it was still caught in the paradigm of high fidelity. In Part IV, I will illustrate the irony of how copyright, which was originally constructed as a mechanism to serve the public interest, has become dominated by private interest. Dramatic irony is a further category, which has a role in retrospective or historical analysis. It unites the author and reader with a superior knowledge (in this case from hindsight) that a character's expressed intentions are doomed, such as the music industry executive cited in Chapter 5 who declares that '*we will never do electronic retailing. We will never allow people to download our music*'.

Narrative and stories: here the interest is in the process of sense-making in organizations through the construction of themes and accounts of events and ideologies, not only by individual authors, but by others within an organization, who engage with the stories and influence the direction they take. Whilst narrative and stories are literary devices which can be used for rhetorical purposes, the focus tends to be more towards sense-making and simplification than persuasion. A concern with the dominance of some stories (masterplots or master-narratives) and the suppression of others often moves narrative analysis critically towards macro-level social and cultural themes. Media coverage of the recorded music industry in recent years reveals examples of competing narratives which construct the causes of the industry's woes with an emphasis on one of two dominant diagnostic plots defined by their main characters: pirates and dinosaurs. The first is the more sympathetic version. It cites rampant piracy, massively stimulated by the widespread availability and convenience of digital copying and file-sharing technologies, as the cause of the 12-year decline. These stories involve fear-mongering visions of an economic apocalypse in the world-leading US and UK creative industries, if intellectual property is not adequately protected by reinforced anti-infringement measures. The alternative story blames the economic and cultural complacency of the historically dominant record companies, i.e. the dinosaurs. Their institutional rigidity and the biases and staleness of their ageing white male executives, who have no idea how to create value for artists and consumers through new business models and services in the digital age, combine to form a recipe for disaster. Fears of, on the one hand, an inexorable rise in the digital theft of music and, on the other, the disintermediation by any one of a large number of new media and technology innovators have caused these content-owning companies to establish highly protective and conservative strategies. Such strategies constrain content usage, leaving content-owners blind to the commercial opportunities provided by new technology. By relaxing these protective strategies and creating more flexible products and services which respond to the consumer's demand for choice and convenience, then surely there must be an opportunity for growth in new business models to compensate for the decline in the physical format model of production and distribution. Or so this story goes. If not, then, like the dinosaurs, a failure to adapt may lead to the extinction of these music giants who once ruled their world.

It is clear from these four domains that the scope of what can be considered 'discourse' is wide, and can attract some considerable diversity in analytical focus. An added complication is that the term discourse is also often used to denote *interpretative repertoires,* a term introduced by Potter and Wetherell (1987) to describe the lexicon of linguistic resources drawn upon to characterize and evaluate actions and events, and to construct subjects, objects and power-relationships, and, in this usage, discourse analysis can cross all four of the domains described above. Discourses in this sense can be technical and functional, such as a healthcare discourse which underpins the structures, hierarchies, processes and relationships of medical practitioners and those needing medical attention, or a corporate governance discourse which establishes directorial behaviour at board meetings. Discourse in the sense of interpretative repertoire can equally refer to larger sociopolitical themes, where they tend to be referred to as 'grand' or 'mega' discourses.

Grand discourses

A grand discourse, or a mega discourse, refers to interpretative repertoires which construct human subjectivity and power relations in a more enduring and institutionalizing way. Here discourse does not refer to the dynamics and devices of micro-level interaction between individuals, but instead to larger phenomenological concepts of social, moral, scientific and economic norms. It is often argued that these discourses have become so convincingly embedded in terms and expressions which dominate the ways that certain things are talked and written about, that they pass for universal truths, or 'common-sense'. Gender relations, political distinctions of left and right, ethnicity and social justice are typical areas of focus amongst those with a critical-theorist approach to the analysis of discourse. The upper case (*D*iscourse) is often used to emphasize that what is being referred to is an established and broadly accepted set of laws, rules, explanations and ideologies which are embedded in language. The concern with the hegemonic tendencies of capitalism, democracy, globalization and neo-liberalism would be examples of where the identification of grand or mega discourses is helpful to the analytic process, even if the very naming of them as grand or mega becomes its own politically loaded construct. In the context of the music industry, discourses of cultural economy, technological determinism, capitalism (intellectual 'property'), the cultural and intellectual commons, and social justice are, amongst others, candidates for being labelled as grand or mega discourses.

When we talk about such things, we generally use terms provided by history. These themes have been the concern of social constructionists and writers who have explored the historical role of language in the constitution of social and psychological life. The most well-known name at this end of the discourse spectrum is Michel Foucault, who was particularly interested in the historical roots of discourse and power relations.

> Let us give the term 'genealogy' to the union of erudite knowledge and local memories which allows us to establish a historical knowledge of the struggles and to make use of this knowledge tactically today
>
> (Foucault 1980, p. 63)

Foucault believed historical aspects had to be carefully explored and stated that, 'unlike the major and often colourful "master" narratives of history, genealogy is gray' (1977, p. 139). This distinction discourages us from being seduced simply by the excitement of disrupting what would otherwise continue to be held as a self-evident truth. His work suggests that the process through which an account can be given of the historical emergence and political construction of institutional rules and accepted social 'norms' is not neat, satisfying and digestible, but littered with contradictions that require painstaking documentation. In this context, the genealogical reflections in this book are not rigorously Foucauldian, but

I believe they nevertheless help to shed some light on the evolutionary power of discourses which have supported an industrial hegemony.

Are we masters or subjects of discourse?

One of the most important questions raised in this book is to what extent are we masters rather than subjects of discourse? We are sometimes, but not always, conscious of the language we use and how it draws on the power of prevailing discourses. Tactful or 'politically correct' statements may not represent our feelings or convictions, but we may toe the party line because it suits other interests and priorities we may have. Whether we consciously embrace or resist particular discourses, we get caught up in their reproduction. We adopt particular discourses, or counter-discourses, because they are plausible, well-packaged, off-the-shelf products. They make sense to us in ways which are more accessible, and less mentally taxing, than critically re-evaluating all of life's complex dilemmas, and then building our thinking and our arguments from first principles. 'The trouble with words', laments Dennis Potter (2012), 'is that you never know whose mouths they've been in'. By taking cognitive short-cuts with pre-packaged discourse or sound bites, and not really thinking about the next word we use, there is a risk that institutional language subjugates us, as unwitting users, into a default position of uncritical acceptance of the policies and ideologies which are constructed through particular discourses, whether they are social, political, industrial or corporate.

Analytical approach to the texts

Carla Willig, an authority on discourse analysis methodology in psychological research, remarks,

> dominant discourses privilege those versions of social reality which legitimate existing power relations and social structures. Some discourses are so entrenched that it is very difficult to see how we may challenge them. [] At the same time, it is in the nature of language that counter-discourses can, and do, emerge eventually.
>
> (Willig 2001, p. 107)

Discourse analysis which adopts a critical-theorist approach aims to highlight, or to unmask such dominant discourses, not with an agenda of promoting a counter-discourse, but merely to make the social issues more visible and discussible. The researcher must examine both the constructive and functional dimensions of the text. This means identifying not only how subjects and objects are constructed, but also how such constructions vary across different contexts, and with what objectives or consequences they are deployed. In practice, empirical research projects which employ forms of critical discourse analysis to approach questions of industrial strategy are quite rare. The analytical approach for the research in this book is influenced by Norman Fairclough, who is a pioneer in the field of critical discourse analysis, though his work is focused on the language of politics. Nevertheless, he gives a useful analytical framework for critical discourse analysis, broken down into stages (Fairclough 2001), which can be summarized as:

- Focus on a social problem with a semiotic aspect.
- Identify obstacles to the problem being tackled.
- Consider whether the social order 'needs' the problem. Who has an interest in it not being resolved? Does discourse preserve orders, divisions, or authority?
- Identify possible ways past the obstacles. Look for hitherto unrealized possibilities for change in the way social life is currently organized.
- Reflect critically and reflexively on the analysis.

The social problem in the context of this book can be expressed in a number of questions which centre on the custodial dilemma which was expressed in Figure 4.1 in Chapter 4, and which will be revisited in Part IV. That is to say, the problem of copyright in the digital age: to what extent should traditional content-owners and producers retain the right to control what is produced and how it accessed? More fundamentally, what right does an author of an idea, sound or image have to own their work and to restrict the circulation, enjoyment, reproduction or adaptation of their work into new creations? How far can claims to originality go? Is all creative (and scientific) output to some extent derivative, being influenced by and building upon the work of those who went before? What impact does the tightening or loosening of intellectual property laws and enforcement policies have on the economy and on the well-being of society in the digital age? Who wins and who loses from technological innovation, and to what extent should the government interfere to protect or promote old or new stakeholders?

Each of these questions is indicative of competing interests which are not easily reconciled. Any shift of the prevailing balance of power will have winners and losers. The social problem is therefore how to referee this contest in social relations, economics, politics and ideology. The Greek word for contest is *agon*, from which are derived the central characters in storytelling, protagonist and antagonist. I will introduce the various 'agonists' in a number of industry tales shortly, but first I consider the concept of strategy as storytelling.

7
Strategy as Storytelling

If 'storytelling is the preferred 'sensemaking' currency of human relationships among internal and external stakeholders' (Boje 1991, p. 106), then 'surely strategy must rank as one of the most prominent, influential, and costly stories told in organizations' (Barry and Elmes 1997, p. 430).

Strategic management can be seen as a form of fiction, not in the sense of being false, but of being made up. The plausibility of corporate stories is a critical element in the field of strategic sense-making, e.g. not 'what's the story here?' but 'what's a story here?' (Weick 2001, p. 462). Unlike traditional strategizing, which is about fit, prediction and competition, the narrative view is about direction. The strategist-storyteller can choose from a vast array of characterizations, plot lines and themes to explain where or how an organization 'is', how it came to be there, where it might be going and what or who it might encounter on the way. In this exercise, they can attribute many things including motive, blame, credit, cause, agency and significance.

Corporate strategizing is heavily influenced by the management consulting industry. The fee-driven project culture of strategic consulting requires that narratives must never become stale or tired. Thus, strategy-making is in a perpetual cycle of re-invention to create at least the appearance of innovation and activity. This does not necessarily mean an evolution or sense of progressive improvement, but it is necessary because tastes and preferences change amongst organizational stakeholders, and in society at large where there is an insatiable appetite for novelty. From the late 20th century onwards, a dominant business culture has assumed the existence of a highly competitive landscape which is increasingly complex and subject to ever more rapid and fluid change through technology and innovation. It therefore tends to attach a very high premium to 'latest thinking'. Discourses born of the internet, globalization and environmental concerns are just three examples of why strategy has had to adapt its narratives to remain fresh, even if the underlying principles sometimes remain the same. New plots and characters are actually quite rare, but strategists can vary the weight of importance amongst certain themes or characters; for example, shifting the role of central antagonist from competitors to disruptive technologies, to regulators or to anarchists.

The strategic landscape of the music industry

For the 50 years up to the end of the last millennium, the business model of the large multi-national music companies had remained remarkably stable, in spite of innovations in sound reproduction. Strategy was not prominent, either as an organizational topic or as a formally resourced function. Strategic differentiators between the big players were few; examples include the approach to growth (in-house talent development versus the acquisition of established artists or catalogues), or the degree of investment in *local* repertoire, meaning not simply relying on selling *Anglo-American* hits in every country around the world. There were also some tactical options along the way: the extent to which one should control certain elements of production and distribution, such as studio recording or the manufacturing of physical products (vinyl, cassette, CD, DVD). Nevertheless, the strategy had been broadly the same for all major players, with the result that the intellectual and management focus of a whole industry has been contained quite narrowly within the highly competitive hit-driven domain of artist talent discovery and development, and subsequent promotion and distribution.

Before considering what went awry with the strategic processes in the new millennium, I would like to reflect on the definition of strategy.

What is strategy?

> Strategy *n*...the art of a commander in chief;...the art or skill of careful planning towards an advantage or a desired end....
> (Oxford English Dictionary)

A notable element of this definition is the repetition of the word 'art', which may provide an explanation of why, after decades of attempting to legitimize strategy both as a science and a profession, there are still widely divergent views of what it is, and in particular what strategists actually do. With reference to the literature, Hendry (2000) lists the following diverse characterizations of strategy:

A text or document; a set of ideas; ambitions or intentions; shared cognitive schema; an analytical process; a management process; a sequence of investments; a pattern of events; a pattern of products and markets; a set of relationships. (p. 969)

When strategy can be alternatively an idea, an object or a process, it is not surprising that attempts to capture its essence can lead to Balogun et al.'s (2003) catch-all definition as 'what is done and what is not done' (p. 199). Whilst logical, this definition is not practically helpful. It is too broadly inclusive, making it synonymous with 'management' and illustrative of the kind of non-committal mystique which surrounds strategy and suggests involvement in everything, and responsibility for nothing. Rumelt (2011) states that 'the first natural advantage of good strategy arises because other organizations often don't have one' (p. 10). They may think they have one, but this is because too many people have become accustomed to equating strategy with ambition, leadership, vision, exhortation, planning and budgeting, or the analysis of competition. Whilst these are important organizational requirements, it is not helpful to confuse them with strategy as the output often becomes diluted, conflicted, incoherent, or just bland.

Given the lack of a clear front-runner in strategy definitions, I may as well add my own. My proposal focuses on the process rather than the output, and draws out two quite different elements. Strategy is:

any activity involving the **perception of change**, whether experienced or anticipated, whether desired or feared, and the consequent formulation of a coherent and **actionable response** to such change.

The response may be to do nothing. Thus, logically, staying the same could be described as a strategy, though in my experience, one with which few strategy consultants would be comfortable. In the context of the recorded music industry pre-1999, where broadly 'staying the same' extended over many years, such a strategy removed the need to have a strategist as a formally and explicitly described member of the executive team. Post 1999, both of the record companies I worked in, Universal and EMI, experimented with high-level strategy executives, though in neither company did anyone hold the role for long. In the latter, consecutive strategy heads were replaced twice in less than two years prior to the CEO resuming direct responsibility for the function. At both EMI and Universal, I held positions which meant that I was involved in many of the high-level strategic discussions. It was the unsatisfactory nature of strategic processes, despite the high intelligence of participants, which led to my original research, and which was sponsored by EMI.

To understand my approach, it is worth breaking down my definition of strategy into its two components: *perception of change* and *formulation of response*. There is a mountain of literature on the latter, but relatively little on the former; i.e. the question of the constitution of the change problem itself: *how* and *by whom* are relevant changes perceived, identified, investigated, defined and prioritized? This imbalance suggests that the perception and selection of changes which challenge organizations, usually articulated as threats and opportunities, are relatively uncontested, and that the majority of the intellectual debate surrounds the strategic responses. Such an inference is problematic and ignores the question of whether the opportunities and threats facing the organization are constructed *independently* of the strategic process. Pettigrew (1977) raised the political agenda questions of which dilemmas to promote and which to suppress, who promotes them, how do they generate demand for their view and mobilize power in support? Knights and Morgan (1991) went further in suggesting that strategy tends to constitute only those change problems which it proposes to solve from within its own pre-defined repertoire of strategic discourse.

By way of background to this distinction between the cognitive aspects of change (the perceived need for strategy) and the associated response (the formulation of strategy), it is helpful to start by looking at the two broadly alternative academic positions on strategy. These categorizations are highly generic simplifications and I am conscious that such a polarized approach can deceive as well as clarify. The first position, which may be called rational (Hendry 2000), is characterized by the orthodoxy generated by much of the literature in which strategy is perceived as a set of rational and largely prescriptive economic techniques for identifying strategic choices and making decisions in a complex and changing, but essentially objective and knowable, external environment (Ansoff 1965; Schendel and Hofer 1979; Porter 1985).

The alternative position, or group of positions, is the challenge to this rational-intentional orthodoxy by those who accept the limitations and biases of human endeavours in pursuit of the rational (Simon 1957; Cyert and March 1963; Pettigrew 1992; Mintzberg 1994). In an unknowable and largely socially constructed world, the focus is more on the cognitive, contextual, process and political nature of strategy practices. These may be categorized as 'action', 'learning' or 'interpretive' perspectives. Through conceptualization of strategy as an emergent social process, these alternative positions contribute to an understanding of the limitations of the rational approach. Emphasis on the primacy of strategic choice and decision-making may lead to only a partial and disconnected understanding of the larger social process of strategy, which often fails to take into account that strategizing occurs both before and after strategic choices are formulated and decisions made. It is an ongoing process which is a continual struggle to make sense of a changing environment, and then to construct provisional business solutions which can hold firm as long as possible, or until something more productive can be achieved based on new knowledge.

A number of scholars have documented the broad scope of the field of strategy. Mintzberg et al.'s (1998) *Strategy Safari* critically traces the evolution of ten prescriptive and descriptive schools of strategy over the second half of the 20th century. De Wit and Meyer's (2005) *Strategy Synthesis* takes a more dialectical approach to the taxonomy of strategy, highlighting the most common paradoxes or debates within the field. Both illustrate the ambition and range of the field, which is to determine the how, what, where, when and why of thinking and acting with reference to organizational goals.

Strategy as sense-making

My original concern with strategy was more narrowly focused on the cognitive aspects of strategizing. By 'cognitive', I refer to my interest in the process of how people perceive, and come to feel that they know things; how they make sense of things, and how such processes can both provide and limit available courses of action. In this context, I am interested in the competing perceptions and narratives surrounding the threats and opportunities which face an organization and which reproduce particular constructs of the industrial world and of cultural production. In my experience, Kenneth Andrews' (1971) analytic framework for identifying strengths, weaknesses, opportunities and threats, or 'SWOT analysis' as it is popularly known, is still the most common component of any corporate strategy process, and its familiar two-by-two grid can be found in most PowerPoint strategy documents.

A SWOT analysis invites a narrative interpretation of the world, especially in the editorial decision to include or exclude particular threats and opportunities. The outcome of a SWOT analysis presentation will inevitably involve the corporation as hero-protagonist overcoming its weaknesses, battling with threats, and using its strengths to exploit opportunities. In many outcomes, especially where a corporation has a large share of an existing market, the strategy is to shape the market according to the strengths and the will of the corporation. This certainly seemed to be the case for the first half of my career in the music industry, with executives being quietly very certain and secure about their businesses' leading place in a stable and well-controlled value chain. It seemed this value chain would endure forever, based on a common understanding of 'natural' rights and roles in cultural production. In the pre-1999 world, stories and strategies did not need to be loudly and explicitly proclaimed because the discourse which underpinned them was largely consensual and therefore dominant.

By 2001, that confidence was fundamentally shaken by an alternative discourse which questioned music companies' rights to control music production and dissemination. Corporate duty forced executives to reinforce the old strategies and stories, because their balance sheets and share prices were under serious threat. As market leader, Universal was in a strong position, having integrated PolyGram prior to the arrival of Napster, and post-merger synergies delivered shareholder value, giving them breathing space to adapt their cost structures and to build even more market share. The merger of Sony and BMG was based on the same medium term strategy: more or less business as usual, but with better scale economies. EMI and Warner were more precarious, especially once their attempted 2006 merger failed.[1] Having regular external reporting obligations in a shrinking market made it extremely difficult to find a compelling story, one which would persuade investors to hold onto their shares in the long term and ride out the storm. Privately, the more open-minded and reflective executives and managers were grappling with the implications of the changing environment, and with how they might construct radical new strategic stories to tell which would be sufficiently plausible and robust to withstand the potentially devastating new consumer behaviour. At EMI, these ideas never saw the light of day, and the strategy remained, once again, a story of 'business as usual'. In late 2006, I attended the EMI board meeting at which the conclusion was reached that even the most optimistic of projections, upon which our strategy was based, could not deliver as much value for shareholders as was being offered by the private equity groups which were targeting us. We logically had to recommend the sale of the company as the best option for shareholders, knowing that it would mean dismissal for most of the executive team. Nine months later the company was sold, on the eve of the global financial crisis. At the time, and also with hindsight, this outcome was the best deal for shareholders, though it was rather disastrous for the acquirer, Terrra Firma. They were caught out not only by the disastrous timing, but by their own lack of strategic solutions, assuming merely that further cost-cutting, and hiring of high-profile 'new media' executives would deliver value from the deal. Terra Firma defaulted on its debt covenants and lost control of EMI to Citigroup in 2011, after which EMI was broken up and sold to competing record companies and publishers.

I mention the fate of EMI with some regret, because despite the avoidance of further share-price decline, which public shareholders would undoubtedly have suffered had the sale in 2007 not been recommended, there was still a sense of failure that management had not been able to develop a strategy which could have avoided the sale of the company. My research was stimulated, to a large extent, by observations of the more private musings of highly talented individuals, and the constraints on organizational dialogue which bordered on taboo in some cases. It struck me that it was often difficult for people to have open and unconstrained dialogue on certain themes. For example, it felt personally risky even to consider whether file-sharing might not be such a terrible thing, unless one had a PowerPoint deck illustrating how it could be brought under control and monetized. A similar sensitivity existed on the topic of copy-protection technology on CDs. The research conversations for this book thus represented a relatively safe space where participants could, to a greater or lesser extent, test out the credibility of their latest thinking on complex questions of business models, strategy, and the fairness and effectiveness of the processes of cultural production in the 21st century. Although all participants were highly intelligent, articulate, confident and persuasive people, at times it was apparent that they were feeling their way, rehearsing new phrases to test their plausibility.

This practice of clarifying one's thinking out loud is reminiscent of the phrase 'how can I know what I think, until I see what I say', a concept which is generally attributed to Graham Wallas (1926) or E.M. Forster (1927). It captures the idea that, in a largely unknowable and unpredictable world, the best we can do is to experiment with provisional, plausible arrangements of words, which may be sufficient until something better comes along. This way of approaching social questions assumes that all knowledge is socially constructed. That is to say that language does not merely describe or represent reality; language actually constructs reality. Words come first, tentatively negotiated to test their efficacy; structures and behaviours come afterwards, eventually becoming institutionalized as social reality, if they prove to be sufficiently robust. If you change the words which we from time to time mutually agree to describe things, then the 'reality' changes also. The history of copyright from Gutenberg onwards provides rich examples of such changing reality, and I will address that further in Part IV.

In the meantime, I do not want to become distracted by the abstract epistemological questions raised by social constructionism, as interesting as they are. It is, however, useful to propose that in the context of the music business, a social constructionist view argues that there are no industrial nor organizational realities beyond the words which we use to describe them. When concepts are new, fragile or fast-changing, this is not a difficult concept to digest. When it implies the disintegration of a century-old industry with powerful and influential stakeholders, or the modification of author 'rights' which were written into statute so long ago that the original intent has been forgotten, it is, not surprisingly, more difficult to accept. The evolution of new realities is thus constrained by the pace at which new voices can influence the adoption of new or modified discourses which shift the balance of power between old and new stakeholders. In this way, narratives may equally serve to emancipate and enlighten as to oppress and exploit (Gabriel 2004).

In the next chapter, I start to analyse the transcripts of the stakeholder conversations in order to disentangle, test and qualify the multiple claims on meaning which strategic storytellers make. In doing so, I will demonstrate how alternative perceptions of change are constructed in order to influence the narrative agenda.

8
Identification of Key Constructs

The first steps of analysis of the original conversations led me to identify a number of discursive objects which had some bearing on the research. The four most relevant and recurrent were:

- music
- the consumer
- the record company
- technology

Each of these four discursive objects was constructed in a number of different ways which were indicative of alternative, sometimes competing, views, not only of the objects themselves, but also indicative of broader views of music as a cultural industry. Figure 8.1 summarizes the alternative constructs:

Music as:

- Artefact/product (to be owned/collected)
- Utility/service
- Honey trap
- Social/cultural capital

The consumer as:

- King (to be served)
- Subject (to be led or empowered)
- Community member
- Thief
- Patron (through active fan support)

The record company as:

- Patron
- Filter (homogenizer/pluralizer)
- Custodian
- Factory
- Bank

Technology as:

- Progress (an irresistible force)
- Charlatan (over-promised/under-delivered)
- Enabler/friend
- Disrupter/foe
- Plumbing ('mere conduit')

Figure 8.1   Object constructs

Music

References to music as an *artefact*, i.e. as a manufactured and artistically packaged *product* to be bought, owned and kept as part of a collection, were common. The ontological separation which gives recorded music output a separate identity from its source input is an important distinction. In this conception of music, the recording studio is a place where the raw materials are assembled and refined. The end product, i.e. the recording, is not a representation or copy of something else. It is the culmination of a creative process, *the* original, the 'master recording'. Given the age range of the participants (early 30s to early 60s), this is not surprising, as this has been the dominant product conception for the music industry for many decades, but it is interesting that this conception of music is losing its primacy as a driver of value, especially as the appetite for live music increases.

One of the older participants felt that the format had lost some of its artefact value with the demise of the vinyl LP, though another robustly defended the CD as something which can still be beautifully packaged. Some point to its durability and collectability:

> If you examined people moving home, what…most people take their music collection. Very few people throw out their music collection when they're moving home. You know, you play things you paid £10 for 20 years ago. It's unbelievable value for money. After your latest jeans have been thrown away, your shoes are worn out, it doesn't wear out.
>
> (Chairman IFPI)

and others to its suitability as a gift:

> it still makes quite a nice little product. It's rather difficult to…I can't imagine…I certainly didn't get fired up by being given a voucher for iTunes. It's not the same as being given a disc is it?
>
> (Head of strategy, Orange)

Most references to music as an artefact or product were positive statements. Occasionally though, the disc format is referred to in a way which is indicative of contempt that certain people can only conceive of music that way:

> at the end of the day, the guy who said this in the EMI meeting was strutting his stuff, because he was the head of something that's plastic with a hole in the middle. That was his job.
>
> (CEO Commercial digital radio)

The 'guy' in question is an un-named record company executive, who in 1998 said that 'we will never do electronic retailing', his myopia being a clear object of ridicule to this participant.

In stark contrast to its construction as an artefact, music is also described as a *service* or *utility*. This is sometimes explicit, as we saw in the many references to the subscription or flat-fee licensing models, but also implicit, such as by references to music being delivered by 'pipes', or by reference to its intangible and ubiquitous properties. Radio broadcast and internet streaming services also contribute to this insubstantial and ephemeral quality of music. Unsurprisingly, the participants who most regularly describe music in non-physical terms are those whose commercial interests derive no value from the physical product, such as those with stakes in radio, mobile phones, and the provision of internet, social networking and digital music services. These are, for the most part, the same stakeholders who, along with those with an interest in selling digital music players, are accused of strategically conceiving music as a *honey trap*. In the case of Apple, music is a way to sell more iPods; for mobile phone companies, music is, along with news, sport, video and 'adult' material, just another form of 'content' helping to promote higher 3G and 4G tariffs and more profitable handsets.

Music as *social capital* is a common construct of all participants. For Nokia, Orange and MySpace, the social value of music is explicitly at the heart of their business models:

> a Finnish guy I used to work with at Nokia. He sort of made that point quite strongly which is that communities don't exist without social objects around which they can perform and it can be communities around music
>
> (Corporate strategy manager, Nokia)

> the idea of building your music store linked to your phone as something that is at the heart of your social network. You know, there is a role for a mobile operator in that.
>
> (Head of strategy, Orange)

> MySpace music, whilst it has a great importance in music, most of what people are doing on MySpace is social networking and then kind of consuming content second.
>
> (SVP marketing MySpace)

> It's [music on mobile phones] in the school ground, you know, it's right there. People say, 'I've never heard that' and you say, 'I've got it'.
>
> (New media financier)

The social value is also recognized by others not directly involved in new media:

> They're able to pass it on to their friends quicker than they've ever been able to and I think capturing that excitement is a great thing.
>
> (Artist manager)

> You talked about what you heard on the radio. And now it's much harder to talk about something that you have in common because everybody has gone off in individual ways. You know, which is why there's people in social networking sites . . . is the new glue that binds people together.
>
> (CEO, commercial digital radio)

and the social capital of attending concerts and music festivals is seen to have become stronger than ever:

> I think they like all going to something together with other people and being at one with a live performance at . . . they just . . . it fits the zeitgeist if you like.
>
> (Independent A&R/producer)

Loosely connected to the construct of social capital is *cultural capital*. The concept of cultural capital is a complex one. Together, social and cultural capital are terms which describe the empowering and privileging properties of non-economic resources, or of resources which are difficult to measure in accounting terms. At a consumer level, music is seen as something which gives people an empowering or confidence-building identity:

> You want to own it. You want to be part of it.
>
> (Independent A&R/producer)

This craving for identity is seen by some as an opportunity for exploitation:

> how can you use music to really start to capitalize at the lower end of the market, where it is clear that they [*young music consumers*] define . . . a core part of what they are is the music they listen to.
>
> (Head of strategy Orange)

Though difficult to measure, corporations do assess the value of symbolic capital, as in the case of Starbucks being able to offer Paul McCartney a more lucrative deal than EMI (his long-term label) could:

I think Starbucks had two elements in the offer they made to McCartney. One was the actual record sales but the other was the image and hope people would see Starbucks as being cool. Because the media view of McCartney is a lot cooler than the consumer view of McCartney.

(Former CEO, EMI)

As we saw in the section on cultural custody, there is a tension between those that want to hold on to the system of cultural intermediation, and those that see commercial opportunities in the liberation of the prevailing systems which legitimize cultural and symbolic capital. MySpace was one of those:

MySpace really now is about people, content and culture and the intersection of that. The evolution of that as a business is what remains to be seen. I mean if anything you were going to ask the oracle is, how do people, content and culture come together through community, a networked future?

(SVP Marketing, MySpace)

There is much one might further explore on the subject of cultural capital but for now I'll leave the final word to the government minister, explaining how the Prime Minister had set his objectives for growth in the soft resources of the cultural economy:

He [*the Prime Minister*] said it should appeal to me because he knew from things that I'd said, and he'd know...and that I'd know that he believed, that the areas of the economy which were likely to grow and grow fastest were the areas where innovation, invention, creativity are central. And he reminded me that I'd made a speech which hadn't gone down wonderfully well at the TUC conference in the past, where I'd said we probably aren't going to beat on metal and dig things out of the ground that much and make a living out of it in the future. So he said, 'make that part of the economy grow'.

(Government minister)

The record company

By now, many of these alternative constructs will be familiar to the reader, so I will reduce the number of text examples for the sake of pace.

The constructs of the record company interconnect with the other three object constructs in Figure 8.1. So, for example, those who view the record company as playing a key role as *patron* would tend to see music as *cultural capital,* and might seek to protect artists from those who see music as a *honey trap,* or those who would devalue the artefact by treating music as a *utility.* The record company as *filter* is a common construct, though this relatively neutral term can cover both positive and negative depictions. On the positive side is the role of cultural intermediary, constructed as a discerning authority exercising cultural patronage, which might in some cases have pluralizing effects as music is developed in genres. On the negative side, others view the record company filtration process as elitist and homogenizing, privileging the narrow demographics of its executives, as we saw in the depiction of 'middle-class public school-boys playing guitars'. Music as *cultural capital* would also be the resonant construct for these kinds of comments.

The record company as *custodian* has both a physical and a cultural dimension: the former connecting to the construct of music as *artefact* to be protected against theft; and the latter to music as *cultural capital* to be protected against falling standards of art and taste in a country which takes pride in its musical legacy and in the ability of the UK music economy to punch above its weight internationally. These concepts were amply demonstrated in the earlier section on custodial tensions.

The record company as *factory* is linked with the construct of music as *artefact,* and is in clear contrast to music as *service* or *utility.* It is part of a wider depiction of the physicality of manufacturing and distribution infrastructure, which at its peak was producing billions of discs globally. Unlike the construction of *artefact,* which on balance was more positive than negative, the factory depiction in these conversations is negative, as seen in the rather disparaging remark about the 'sausage factories' and 'biscuit factories'. As record companies have now divested themselves of their factories, there was also a sense of history or obsolescence regarding this construct.

The clearest constructs of the record company as a *bank* are connected with patronage, and came from the independent A&R/producer, the artist manager and, as illustrated here, the composer/producer

But for the guy who's starting out who's really passionate, where does he get the money from to buy the recording gear? Where does he get the money from to keep him going while he's, you know, getting good at his craft, learning to play instruments and you can't … I don't think you can do it without that sort of money. But it's all very well saying music should be free but who's going to pay for it to be made? That's what I don't get.

This remark was prompted not so much by discussion of the economic downturn in the industry, as by irritation at the practice of superstar artists, such as Prince, of giving their new recordings away for free. It suggests that artists need a special kind of funding, and in this sense record companies are seen as specialist providers of capital.

There are other constructs of record companies. For example, the record company as *studio* focuses on the discourse which puts the recording process at the heart of the industry's activities and value contribution. It connects with a construct of music as *recorded sound*, and with another, now obsolete but important, construct: that of the record company as a *laboratory* or technology innovator. These have implications for the important distinction between *music* companies and *recording* companies.

The consumer

Alternative conceptions of the consumer can similarly be mapped to the themes emerging from the other object constructs. The most common descriptions of the consumer refer to the state of their agency or empowerment. *Subject* refers to instances where the consumer is someone to be led or directed in their consumption and tastes or, in more extreme cases, exploited or even 'fucked' (an accusation directed towards Apple from a participant who did not wish to have this remark attributed to him/her). Often, the subject construct is implied in the context of participants explaining how or why consumers need to be more empowered. The construct as *king* reflects the high number of references to the customer as someone who must be listened to, served, or who needs to be in control:

By ignoring the consumer, that was not containable, because the consumer's always right. […] they are web surfers and what this web surfing means, it means you are in charge of your destiny.

(Former CEO, EMI)

accept the fact that people are not going to do what we want them to do in this space.

(Head of interactive music, BBC)

Whether positive or negative, these empowerment references are linked to alternative views within the constructs of music as social, cultural and symbolic capital, and of record companies as filters and custodians. The construct of the consumer as *community member* is clearly a favourite of MySpace, Nokia and Orange. Construction of the consumer as *thief* is indicative of where a participant sees the primary custodial responsibilities of the recording industry:

> You can see that young people are happy to pirate and aren't that frightened…

> […] Do you think people get old and stop cheating?

> […] We had a number of focus groups with sixteen year old kids. I think people were quite tempted to just lock the door and beat them up – [*Our head of sales*] just wanted to get at them with a stick.

These three remarks were made by the head of digital at a major record company, and illustrate the ways some record company personnel feel towards some music consumers. The next comment was made in reference to the Radiohead honesty box experiment:

> the terribly sad thing was that so many of their fans and so many of the consumers still chose to steal it, for want of a better word.
>
> (Chairman, IFPI)

Given the legitimate option of paying nothing, the choice of the word 'steal' betrays an entrenched view of consumers who are not prepared to pay for music. This view sees the experiment as a failure, but an alternative 'glass half-full' view expressed surprise that so many paid something, even though they didn't have to. This introduces the depiction of the consumer as someone who actually enjoys financially supporting a band, i.e. the consumer as *patron,* which cross-refers to the construct of music as cultural capital. Sellaband, ArtistShare and Kickstarter are examples of crowd-funding business models which have had some degree of success. Whilst these seem radical and innovative, they are reminiscent of the same spirit of amateur patronage as the 'society' subscription-funded recordings introduced by HMV in the 1930s, and of the subscription models to fund book prints in the 17th century. On a less formally financial basis, the increasing use of social media by active fans to mobilize support for new artists is further evidence of consumers having a different kind of relationship with the production of music.

Technology

At times, technology is depicted deterministically, or even teleologically, as the inexorable march of *progress*. Alternatively, it is a set of tools in the service of organizational managers or consumers, where it may be seen as an *enabler*, and therefore a friend; or as a *disrupter* and foe. Occasionally, it is accused of over-promising and under-delivering, for which, for want of a better word, I have referred to technology as *charlatan*. Finally, because of the numerous references to pipes, and the dominant construct of technology as relating to that part of technology which supports internet and mobile communication networks, there is a construct which could be identified with reference to technology as *plumbing*. The European 'mere conduit' legal defence, available to communications providers to avoid liability for content transmitted via their networks, is an example of this rather passive construct of technology. I give no textual examples here of such technology constructs because they will feature in Chapter 10.

Having identified a number of what I believe to be the most relevant constructs in the context of the research problem, in the next chapter I go on to show how these constructs are part of interpretive repertoires which indicate different views of the world of cultural production by locating them in a narrative framework.

9
A Narrative World

Previous chapters illustrate the themes which I have identified as being of most concern to the research participants, and the different ways in which particular objects relevant to the music industry are constructed. In the process of extracting them and presenting them in abstract form, logically sequenced for manageable appraisal, something is lost. That 'something' may best be described as the messy struggle to make sense of a complex world. It is both a private struggle and a social contest, though the private element is more difficult to capture. The usual demands made of executives for simple rhetorical coherence mean that intra- and inter-organizational encounters are generally the domain of the *social* sense-making contest rather than the *private* cerebral struggle, with the latter only rarely being publicly acknowledged. The research conversations allowed participants a more balanced exploration of the dilemmas and paradoxes inherent in cultural production. Whilst some were more confident in expressing their expertise and wisdom (or ignorance) than others, and despite some evidence of toeing corporate lines, most participants demonstrated an appetite both for reflection and the capacity to be reflexive. This ought to be a desirable outcome, but it does lead to participant inconsistency and presents a challenge for the researcher in pursuit of a clear distillation of findings.

To meet this challenge, and holding to a broadly linguistic methodology, I have attempted to illustrate the plurivocal aspects of the conversations by presenting them as stories in a plausibility contest with each other. A concern with the plausibility rather than the accuracy of narratives and stories is an important element in the field of sensemaking and strategy, e.g., as cited in Chapter 7, not 'what's the story here?' but 'what's a story here?' (Weick 2001, p. 462). The focus is not on finding some kind of universal truth, but on something which is fit for purpose, or good enough. Stories are as capable of being 'vehicles of contestation, opposition and oppression' as they are of 'enlightenment and understanding' (Gabriel 2004, p. 62). There are several properties of storytelling which are helpful in understanding the dynamics of social and industrial change.

One is attributed by Taylor and Cooren (2006) to Algirdas Greimas (1987), and is the facility whereby two chronicles describing two actors following two paths in pursuit of the same object, valued differently, can be subsumed into a single narrative interplay, but with the paths given unequal status. The protagonist's status is euphoric, and the antagonist's is dysphoric. All sense-making presents us with a confrontational logic which Taylor and Cooren express as 'worldview': 'For every story (and every interpretation) there is an anti-story, not just another story [...] It is dichotomous: two-sided, not multi-sided' (p. 119). A story draws its energy from the mutual interference of the two sides. Ultimately, the privileged interpretation of the protagonist prevails with more authority from having credibly recounted the contest between the two sides. In the context of the research conversations, the degree of interference, i.e. the degree of voice given to the antagonist, can be used to gauge a participant's confidence that the protagonist's interpretation is the more compelling one. A sympathetically drawn antagonist might be indicative of a tolerance for counter-narratives, and an acknowledgement that the balance of power in the cultural economy is shifting.

Other generic properties of storytelling include its use of poetic tropes and licence. The latter is described by Gabriel (2004) as a psychological contract between the storyteller and audience. It allows the storyteller to 'mould' (p. 64) the material in the name of giving a voice to experience, consistent with the idea that the truth of a story lies in its meaning, not in its accuracy. Moulding can involve the freedom to attribute many things in the name of plausibility and meaning, such as motive, causal connections, blame and credit, and fixed and consistent qualities to simplify otherwise inconsistent and confusing characters or whole classes of people. Recognizing the inherent risks of poetic licence, researchers are urged not to treat all stories or voices of experience as equally valid and worthy of attention:

> Disentangling these voices, understanding them, comparing them, privileging those which deserve to be privileged and silencing those which deserve to be silenced, questioning them, testing them, and qualifying them – these seem to me to be essential judging qualities that mark research into storytelling and narratives as something different from the acts of storytelling and narration themselves
>
> (Gabriel 2004, p. 74)

With these cautionary principles in mind, I will now present the tales implicit in the participant conversations. Though I have resisted the 'silencing' of any voices, I believe that I have presented them in a way which lends itself to a critical appraisal of the strategic implications of the narratives. I have chosen to refer to them as 'tales', in reverence of Chaucer's masterful construction of a storytelling contest in which each tale takes the name of its protagonist in order to present a holistic social critique. I recognize that the word 'tale' does not have precise significance in narrative theory. In that domain, the tales would be better described as masterplots or master-narratives, being the skeletal story structures which, in one form or another, get reworked in society and which replay common anxieties and reinforce deeply held values and wishes.

Seven tales

Considerable reflection and iterative readings of the transcribed conversations led me to the conclusion that most arguments made by research participants in their diagnoses and prognoses can be reduced to seven tales designated by their central protagonists, namely: patron, discerning-curator, protective-curator, liberal-curator, disseminator, inventor and businessman. The first three of these tales are embedded in the traditional recorded music industry. The next two are highly contested between old and new stakeholders. The last two are not exclusive to the music business, but are well recognized subplots which support or undermine the first five, depending on context. As the tales have some scope for variation, they are best described with reference not only to their protagonists but also to their antagonists, who are instrumental in defining the significance and authority of each tale. Plot headline and protagonist and antagonist variations are as follows:

The patron's tale

Plot: noble patron *discovers* lowly but talented musician, *nurtures* his gift, and fulfils his potential to bring beautiful music to the world.

Protagonist variations: the promoter, the defender, the nurturer, the enabler.

Antagonists: the unforgiving public, the brutal competition, the insensitive capitalist machine (sausage & biscuit factory), the dangers of premature success, the svengali, the allure but false promise of democratizing technology.

The discerning-curator's tale

Plot: discerning-curator makes exciting *selection* of new music, further consolidating position as leading *authority* on musical culture.

Protagonist variations: the talent-spotter, the cultural-guardian, the critic.

Antagonists: democratizing technology, undiscerning disseminators, revolutionary liberal-curators.

The protective-curator's tale

Plot: protective-curator *safeguards* value of music through improved security.

Protagonist variations: the gatekeeper, the custodian.

Antagonists: thieves, pirates, revolutionary liberal-curators, honey-trap setters, anarchists, disruptive inventors.

The liberal-curator's tale

Plot: liberal-curator increases public *awareness* and individual *relevance* of vast music library through improved *accessibility*.

Protagonist variations: the liberator, the democratizer, the concierge, the matchmaker, the pollster.

Antagonists: over-zealous protective-curators, fat and complacent businessmen, elitist discerning-curators.

The disseminator's tale

Plot: disseminator *connects* people with music and musicians, and through music, with each other.

Old* protagonists: the broadcaster, the distributor, the impresario, the shopkeeper.

New* protagonists: the internet service provider, the mobile phone operator, the social-networking service provider, the digital music service-provider, the file-sharer.

Antagonists: over-zealous protective-curators, meddling referees.

*Old and new disseminators are antagonists to each other as their success is mostly defined by their dominance or 'reach'.

The inventor's tale

Plot: inventor makes technology break-through; *improves* lives.

Protagonist variations: the innovator, the visionary, the entrepreneur, the technologist, the scientist.

Antagonists: Luddites, the technologically ignorant or gullible, record company executives who are fat, complacent, slow, unresponsive, and stuck in an obsolete product conception.

The businessman's tale

Plot: businessman *perceives* opportunity in the market, *coordinates* resources to exploit strengths and defeat threats, *makes return* for shareholders.

Protagonist variations: the entrepreneur, the chief executive, the financier, the value-builder.

Antagonists: thieves, pirates, revolutionary liberal-curators, Luddites, short-term investors, disruptive or manipulative inventors.

In addition to these seven tales, which aim to explain the roles and goals of primary participants, there is one, more generic, tale which participants use to claim their own authority and wisdom:

The referee's tale

Plot: referee intervenes to ensure fair and productive play between bakers and slicers in pie competition.

Protagonist variations: the politician, the bridge-builder, the leveller, the judge.

Antagonists: over-zealous protective-curators, revolutionary liberal-curators, thieves, pirates, opportunists, elitists, the unchecked capitalist machine, the stubborn.

As one might expect, this tale is told by the two participants whose representative positions are as a result of democratic process: the government minister and the chairman of the IFPI.

Tin Pan tales and Wiki tales

As already mentioned, some of the tales draw from a traditional recorded music industry discourse, whilst others are products of a more preco-cious web 2.0 discourse of consumer empowerment and liberal plural-ism, where music needs to be set free from those who would cage it. In order to give some emphasis to the competing tales, I have mapped them to these two competing discourses in Figure 9.1. I refer to the former as Tin Pan tales, recalling the old name given to the music industry (Tin-Pan Alley), and the latter as Wiki tales, in reference to the Hawaiian word for 'fast' and adopted by the internet community to describe open format web pages which create widely accessible and

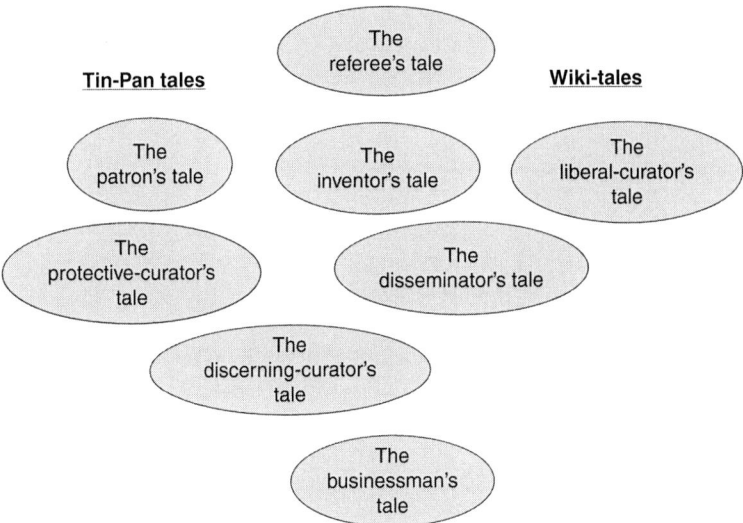

Figure 9.1   A storytelling contest

collaborative community space. Generally speaking, the tales of the businessman, inventor and disseminator are less clearly polarized and are not the exclusive domain of either of these two discourses. However, amongst the research participants, the businessman's tale is more used by tellers of Tin Pan tales, and the inventor's and disseminator's tales are used more by tellers of Wiki tales

It is worth noting that the polarities in Figure 9.1 closely resemble the custodial tensions illustrated in Figure 4.1. Tin Pan tales sustain an economic system of cultural intermediation by protecting cultural and artistic standards, and by promoting the value of expert filtering and nurturing (patron and discerning/protective-curator tales). Wiki tales, by contrast, promote the civil rights of a wider public: to access and to generate cultural capital, by allowing consumers more choice in what they listen to, empowering individuals to become artists outside the narrow filter of record companies and other cultural intermediaries, and by using the wisdom of crowds to dictate quality and value (liberal-curator and disseminator tales).

Other tales

The assertion that all plots can be reduced to those included in Figure 9.1 exploits to some extent the poetic licence (referred to earlier) of omitting material for the sake of communicating meaning. There were alternative constructions of other characters, such as the musician and the music consumer, which suggest the existence of other tales relating to the production and consumption of music. Amongst these alternative protagonist constructions, the implied tales and antagonists vary widely depending on the particular construction of the protagonist, especially the degree to which he or she is empowered or not. The implied plot lines of these additional tales are mostly interwoven with the seven tales I have highlighted.

The musician protagonist variations include: the artist, the entertainer, the diva, the star, the brand, the slave (through contractual servitude), the self-promoter, and the robbed (by pirates). I deliberately group these as constructions of the *musician* (rather than *artist*) to draw attention to the fact that the more neutral or functional word *musician* is very rarely used in the conversations. The more loaded word *artist* is almost exclusively used by participants to describe the originators of music, which is indicative of their claims to the cultural capital of musicians. Within the research conversations, there are 270 references to 'artist' or 'artists'. This compares with only nine occurrences of 'musician(s)', six of these being in the same conversation with, unsurprisingly, a musician. Though musicians' tales undoubtedly exist in the wider discourse of the cultural economy, they do not form a compelling or coherent master-narrative in the research conversations. As for artists, they play a support role, with limited agency, in each of the more dominant tales of cultural intermediation.

As far as the consumer is concerned, I identified alternative constructions of the consumer as indicated earlier in Figure 8.1. There were hints of tales, especially around the consumer as thief, but none of the participants directly employed the taboo master-narrative of the pirate. I will deal with this important omission in Part IV of the book where I illustrate that deep-rooted and polarized cultural narratives which alternately vilify or romanticize the pirate have held the cultural industries (and parts of the new media and technology industries) in political deadlock for well over ten years.

Multiple positions

All research participants employed more than one tale, so I have attempted to summarize their usage in Figure 9.2. The light grey shading indicates where participants make use of a tale or a protagonist position, even if only indirectly. By contrast, the black blocked shading indicates

	Patron	Inventor	Discerning curator	Protective curator	Liberal curator	Disseminator	Businessman	Referee
Music service provider		2				4	3	1
CEO, EMI	2		3				1	
Chairman IFPI				1			3	2
EVP/CFO, Universal Music				2			1	
Government minister		1		2	3	2		1
Corporate strategy manager, Nokia						2		
Independent producer/A&R	2		1					
SVP Marketing, MySpace					2	1		
CEO, HMV					1	2		
Head of interactive music, BBC						2	1	
Head of digital, record company								
New media financier								
CEO, commercial digital radio								
Head of strategy, Orange								
Artist manager								
Composer/ producer								

Figure 9.2 Tale usage

where the participant directly assumes the protagonist position. Where a participant directly assumes more than one protagonist position, the numbers in the black boxes rank the degree of emphasis or commitment to each position, with '1' being the most assertive. The purpose of this table is merely to demonstrate the overall complexity of sense-making amongst participants, and not to draw any quantitative conclusions.

Further analysis

As the previous chapters have drawn out most of the themes of patrons and curators, I will now go into more depth with the inventor's tale in order to explore a fundamental difference in conception of technology, and the consequences this has, both for sense-making and for future strategic possibilities and obstacles facing the recorded music industry.

10
The Inventor's Tale

Entrepreneurs versus technologists

You know, the whole world is going like this digitally. [...] That's kind of just a reality of how technology is changing how the media business operates.

(SVP Marketing, MySpace)

This chapter introduces two important variations to the inventor's tale, which represent fundamentally competing views of the role of technology in industrial transformation. I use examples to illustrate how the research participants' conceptions of their organizations' products are rooted in these alternative views, particularly with regard to whether those product conceptions are fixed or adaptable. Finally, I explore how perceptions of adaptability affect participants' views of agency and influence over industrial change, and what strategic possibilities for coercion and collaboration are opened or closed by these views.

We saw earlier that behind some competing interpretations and judgements about the evolution of the music business, there was consensus regarding the technological developments which challenged the control enjoyed by the traditional recorded music industry model. These are expressed in Figure 10.1.

Away from value-judgements regarding whether these developments are good or bad, fair or unfair, they are relatively uncontested constructs of what has 'happened' in the environment of recorded music.[1] I refer to all these generically as 'developments', because as soon as one starts to describe them it is difficult to avoid colouring their construction. In this section I am particularly interested in those elements of their construction which indicate the participants' views of the role of technology in industrial transformation.

Primary innovators & exploiters

Microsoft
PC manufacturers
Napster, Kazaa, Limewire, etc.
ISPs, mobile phone companies

Apple
and other digital
music player manufacturers
and digital music retailers

MySpace, YouTube, Yahoo
and many other artist and
consumer services for
discovery & recommendation

Figure 10.1   Technological development

The alternative ways of looking at the role of technology in the history of the music business are thoughtfully described by Timothy Dowd (2006). He divides the prevailing views of technology and industrial transformation into two competing depictions, which he calls atomized and embedded. Atomized depictions treat technology as driven by natural laws, rather than being shaped by contextual factors. Technologies provide optimal solutions to problems, and diffuse easily as individual actors negotiate market opportunities in relative isolation, following universal laws of efficiency and profit. He cites Disco and van der Meulen's (1998) lament that in such a depiction 'technologies develop according to an inner logic...and are therefore more or less impervious to human influence. On this view you can't hurry technology, but neither can you constrain it once its time has come' (p. 4). By contrast, the embedded depiction attributes much greater influence to human agency. Markets are context-dependent and affected by inter-firm relations and state policy. Technology is similarly 'contextually contingent', and market change is prompted by 'successful strategies regarding technology, rather than new technologies *per se*' (Dowd 2006, p. 206).

The embedded and atomistic views resonate with the variations already noted within the inventor's tale, particularly the distinction between two types of protagonist: the technologist-inventor, whose primary focus is the *scientific breakthrough* possibilities of technology, and the entrepreneur-inventor, whose focus is the *product* and its market possibilities. In describing an entrepreneur as an inventor, I am making the point that, for most people, new technology is only recognized once it has been conceived as a product and successfully marketed. This focus on marketing in order to establish a dominant product conception is an element of invention which is often overlooked by technologists. This can be clearly illustrated with reference back to Chapter 2 with two startling examples. The first is Edison, who never successfully marketed his cylinder, which began and ended its life as a dictating machine. The second is MP3, which, despite revolutionizing music consumption, was not conceived by its inventors as a consumer product, and their financial returns have been consequently eclipsed by the products and services which MP3 spawned.

The antagonists of these two versions of the inventor's tale are similarly nuanced. For the technologist-inventor, the antagonist is the Luddite, ignorant, or stubbornly in denial of the inexorable march of technological progress. By comparison, the entrepreneur-inventor believes that technological evolution is more malleable and ductile in the hands of imaginative leadership. The entrepreneur-inventor may utilize the same vocabulary as the technologist-inventor in fighting or dismissing his antagonists, but he is more aware, and sometimes more respectful, of their power, intelligence and motives. He therefore views them as worthy opponents rather than as fools.

The distinction between the two views of how technology transforms industry is illustrated in Figure 10.2.

The distinction could be illustrated by comparing Apple's entrepreneur-inventor (Steve Jobs) with Napster's technologist-inventor (Shawn Fanning). More so than most big corporation CEOs, Jobs was closely associated, if not credited, with Apple's inventions, not least the iPod. Success with regard to this invention is generally attributed more to a meticulous attention to design and marketing (product conception) than to being first to market with new technology. The fact that the iPod has become a synecdoche for all digital music players, despite coming to market four years after its early competitors, would support this view. Apple's success with the iPod bears comparison with Kodak's institutionalization of technology through the roll-film camera between the 1880s and the 1930s, and Munir and Phillips (2005) identify the discursive processes which led to Kodak's success. These same processes are identifiable with Apple's marketing which naturalize, legitimize and domesticate a new product conception, transforming it from being a novelty invention into an indispensable possession for a mass market.

In contrast to Steve Jobs, Shawn Fanning's focus on being the first to invent a universal file-sharing technology was apparently not distracted

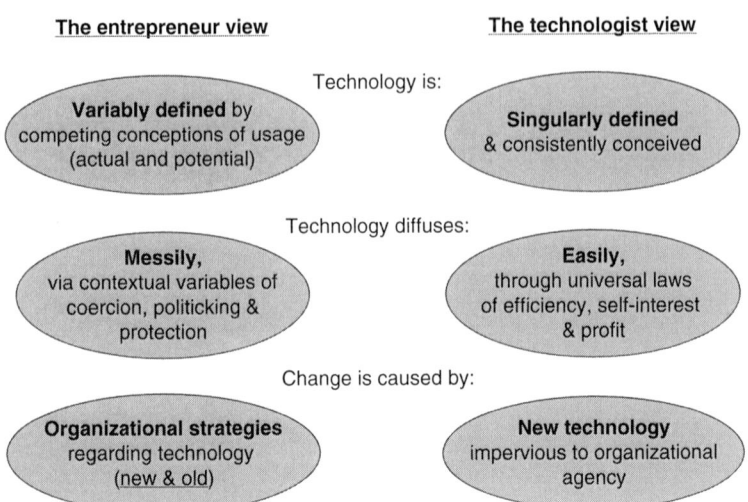

Figure 10.2   The inventor's tale (variations)

by product or market concerns.[2] Spitz and Hunter (2005) refer to the dominant interpretation of the atomized portrayal of Napster's arrival as a 'one-time heroic event' (p. 178), performed by its 'inventor'. They examine how and why certain assumptions about Napster have gained greater currency than others, and argue that the prevailing depiction oversimplifies both the contested properties of the technology and also the social processes which led to Napster being used synecdochically as an organizing principle to build consensus and momentum within a culture already attached to certain values and practices. Rather than being an invention which crystallized in a specific 'eureka' moment, Napster is better conceptualized as a key part of an extended discursive encounter between inventors institutions and interests, leading to the strengthening and legitimization of a 'Web 2.0 Discourse' of collaboration, sharing and community. There are similar accounts which try to attribute the 'invention' of MP3 to Karlheinz Brandenburg, one of the engineers at the Fraunhofer Institute, an attribution Brandenburg is always keen to refute:

> I am certainly not the father of MP3. I know who else contributed in the development of MP3, whose shoulders I stand on, and who else worked on the topic, hence I never refer to myself as the father of MP3.[3]

The very concept of inventing something might seem to contradict the notion that technology is impervious to human agency. However, the urgency inherent in the culture of technologists betrays an underlying determinist view that, 'once its time has come', the only scope for agency conceded by technology is in the race for discovery and accreditation. Gelatt's (1955) account of the apparently unrelated yet simultaneous invention of recorded sound in 1877 by Charles Cros in France and Thomas Edison in the US would support this view.

A participant tells his account of the Napster events from having been involved at a later stage in providing litigation support to the US recording industry:

> Now these [*the Napster*] technology people, I know how they think. They'd go, 'are they [*the record companies*] mad?' Doesn't matter whether they're right or wrong but their thinking will be, 'technology, why on earth wouldn't you do it?' They're sitting there going, 'we've got 18 months [*before someone else invents or discovers it*]. These people [*record companies*] aren't going to do anything in 18 months. Bollocks, we'll do it anyhow'. And it was called Napster, the original peer-to-peer Napster. And it's the frustration...
>
> (Music service provider)

The vignette illustrates the participant's claim (he is himself a direct teller of inventor's tales) that Napster and other file-sharing technologies were not (as often reported) created by young men who were specifically motivated by the desire either to steal music or to liberate music from the tyranny of the record industry. Instead the story is of innovative technologists who were compulsively and competitively driven by the adrenaline of possibility and invention, irrespective of the consequences. The account is preceded by the participant's explanation of the central tension between technology and the recorded music business:

> I was talking to the CEO of PPL[4] and he said, 'I want to know, why do the record people hate technology people, and the technology people hate the music industry? Because I see it every day of the week.' I said, Fran, I'll come and explain it to you. I said, it's called speed and it's called a back catalogue. He said, 'explain'. I said, in the technology world that my background is from, what happens is you invent something, you typically have an 18 month window to exploit it before your competitors can catch you up [...]. I said, the music industry don't have that problem. I said, they have got what's called a back

catalogue. If the music industry is having a bit of a tough time, dig out the greatest hits, dig out this, that and the other and it's called the back catalogue. And the record industry exploits it beautifully. But the IBMs of this world can't have a slightly dodgy year and then say to the market, hey, welcome to the 486, you know, because it doesn't work. So the music industry have got a back catalogue to fall back on if the new products are not available or not performing as well as possible. And therefore that creates a lack of urgency because they always know they've got the comfort value of a back catalogue if necessary. The technology industry don't have that.

(Music service provider)

The back catalogue, as something enduring and worthy of protection, is an important element of the curator's tale, but to the inventor it is a source of both frustration (it gets in the way of invention) and envy or injustice (we don't have one, so why should they?). In terms of intellectual property law, his argument speaks to the perceived discrepancy between a patent (generally 20 years protection) versus the much longer copyright protection for recordings (50 years in the UK) and compositions (author's life plus 70 years in the UK). In any event, according to this participant, even the shorter life of the patent is irrelevant when technology competitors only need 18 months to catch up, suggesting that, in practice, patents only offer a very limited scope of protection.

This is a good illustration of the long length of copyright being a source of complacency and attributable as a diagnosis for the music industry's problems. Back catalogue is something with a 'comfort value', something to 'fall back on' in 'tough times'. The presumption of its enduring quality explains why exclusive control and protection of rights are central strategies of record companies, but such strategies limit opportunities for collaborative dialogue between technology innovators and content-owners. Here, catalogue is a structural concept which has behavioural effects: it breeds a 'lack of urgency' and indecision amongst rights-owners, which provoke impatient technology innovators to act unilaterally, not waiting around for resolution ('bollocks, we'll do it anyhow'). This participant's view, that technology ought to diffuse easily, is not undermined by the obstacles of other stakeholders. Rather, he believes that messy cognitive differences can be resolved simply, if he could just get people together in the same room:

I was talking to Fran and I explained this to him. He said, 'that's fascinating', he said, 'we've got a back catalogue and the IT industry doesn't. Therefore everything that they do has to be faster.' And it's as simple as that. It's nothing more complex. [...] I didn't have that blinkered view about, these people are like this and these people are like that so I think I've got a reasonably balanced view between the frustrations that these people feel and the frustrations these people feel. And also if you were sitting there trying to get a deal done you could probably do it understanding both sides of the coin. You know, so if I were sitting there with the PPL in one school and IBM in the other, I know that neither would understand a word that each other's talking about. But I [...] could bridge it and it's not because of any intelligence. It's just that I've been doing it for too long.

(Music service provider)

This participant offers the clearest direct example of what I refer to as the technologist version of the inventor's tale. This is not surprising; he comes from a technology background, which he pointed out on a number of occasions during the conversation. But the inventor's tale is not the exclusive domain of participants who are themselves inventors, and the extent to which all participants draw from either version of the inventor's tale is indicative of where they attribute blame or credit, and of where they see solutions:

the record companies have let technology happen to them, instead of saying, 'okay, this is clearly a trend now, we are going to be proactive and we are going to accept the fact that we have to go with a different business model or accept the fact that people are not going to do what we want them to do in this space'. And they've been slow to react [...] I wonder whether they haven't listened to the audience enough and rather, you know, listened to technologists and listened to, you know, financial advisers and listened to... I wonder whether they've actually really tracked consumer behaviour and consumer opinion.

(Head of interactive music, BBC)

By way of contrast, he credits the BBC with reinventing radio in the digital age. In the old days, radio was a box which emitted scheduled broadcasts which the listener had only one chance to hear. Now, it is 'chopped up, downloaded, and segmented' as a 'multi-media audio experience':

Well, put it this way, our department, our group used to be called Radio Music. We're now called Audio Music. It's about rich, deep, immersive multi-media experiences of which sound, audio is part of it. [...] And it's quite interesting if you think about it as well... when you think about the new digital radio stations the BBC launched, they don't have the word radio in the title. You have BBC 6 Music. You have 1 Extra from the BBC. You have BBC 7, BBC Asian Network. They don't have Radio 1, Radio 2. They are...now this was clearly a clever bit of long term thinking from Jenny Abramsky but you know, and actually a lot of people have argued that it was wrong because people get confused. People still call 6 Music Radio 6, but not having the word radio in it actually I think was the right thing to do because it's saying it's a brand, a digital music brand.

In this context, Jenny Abramsky, who at the time was the Director of Audio and Music, and, according to Wikipedia, 'the most senior woman in the BBC', could be described as an entrepreneur-inventor. She saw the opportunity to use technology not necessarily to re-invent radio, but to reconceive radio in the minds of the public, i.e. converting it from a passive box-bound broadcast experience to a 'rich, deep, immersive, multi-media experience', and one which puts the consumer (re-defined from 'listener') in control by being able to choose the format and timing of their consumption. In this conception of technology, it is clearly subject to human agency, even though Abramsky had to convince 'a lot of people who argued that it was wrong'. The BBC was sufficiently well-resourced to plough its own furrow in the digital field, though this has not been achieved without controversy and opposition from commercial broadcasters. They complained that the BBC was abusing its public service status by investing in what were perceived as more commercial digital investments. Yet despite the BBC's resources and strong position in UK radio, the research participant still feels that:

> we [*the BBC and record companies*] really should be working together on that [*a new music model*] because...I mean still the biggest way that people discover new music and then go on to purchase it is by hearing a song on the radio. Now obviously that may decline from sort of the 92% of times to sort of something less, but it's still you know, way up there. So I would say, you know, that the record industry should see radio, and particularly radio, as a friend and should, you know, again look at it from the consumer's point of view

The participant implies that the BBC has something to teach the record industry in working together to reconceive a music discovery model, and in the process how to use technology to its advantage. It is interesting to compare the BBC radio view with that of the commercial radio sector, which refers to the relationship with music as a forgotten friend:

I think there's a real problem there, in that I think the music industry
has forgotten who its friends are and I think there is a real danger that
that will damage the relationship. [...] It was a good relationship,
sorry horrible word, symbiotic relationship, you know. You knew
who's back...everybody had to scratch each other's back and it kind
of worked and it was great. [...] But I just think that that relationship has broken down somewhat so now there is...you know I detect
an attitude in the radio stations of, you know, 'sod the record companies! What have they ever done for us? Sod PRS,[5] they've having their
money, you know, we'll decide'. And so...and money...the amount
of money involved now I think has probably caused a split. [...]
There are too many people taking too much greed on the way with
the upshot being that if a radio station wanted to do this [*experiment
with radio/mobile subscription model*] as a source of revenue I think
they'd make something like 4p a track, which is a complete waste of
time.

(CEO Commercial digital radio)

This participant expresses enormous frustration at the fact that technological developments have caused the record companies to act in an ever
more restrictive, fearful and non-collaborative way which has destroyed
the previously robust and symbiotic relationship with radio stations.
In response to the question of whether the radio stations could negotiate
with the PRS and PPL in unison, he goes on:

That's the difficulty. I can see commercial radio acting in unison but
they wouldn't want to do it unless the BBC did. Now I'm preach-
ing heresy here and it's purely, you know, speculation and so on as
to what might happen. But I do worry that...every time that I've
seen a renegotiation of the PRS and PPL rights they've become more
restrictive, and I'll give you a digital example. When we started digital radio one of the big things that benefited...that we were able
to say 'digital radio is better' is because we had scrolling text and the
scrolling text would say, 'playing now is this and playing next is this'.
And within two years PRS and PPL said, 'we're not doing that. You
can't say what's coming next. Can't tell people what's coming next!
They might record it and pirate it.' Well, you know, for fuck's sake!
I've heard some bonkers thinking but commercial radio went, 'oh
no, we're scared, we're scared, we're scared. Oh no, we agree to that.'
So now you're not allowed to say on scrolling text what the fuck's
going to be played next. Bonkers. You know it's bonkers. I know it's

bonkers. But some little prat at PRS thought they had won a great victory. And you know...and I'm sorry if I'm being so harsh about this, but I just think that people have lost their sense of reality over this. If somebody wants to pirate music they're not going to do it by recording off the radio for two very good reasons. One is they can get it a damn site easier off the internet. And secondly the DJ talks all over the music. So what have you pirated? You've pirated something that's got a crap beginning and a crap end which you might edit.

The scrolling text referred to in this extract is a modest example of digital technology being able to improve the radio listeners' experience by informing them of what song is about to be broadcast next, but it is indicative of the culture of the organizations set up to centrally collect and negotiate certain record industry rights that any development which potentially increases the possibility of unauthorized copying is treated with deep suspicion. Underlying the participant's point is a view that the record companies just don't have the right understanding and attitude towards technology:

> The technology on the radios is not nailed down because nobody has sat in a room with the record companies and said, how do we want this to work? [...] But you know, you're the first person I've sat down and talked to, who is entertaining a discussion about it. The last time I met a record company was...It was 1998, you know. But that's the point. And there's either a mistrust or there's an assumption that, not an assumption, a lack of knowledge that actually something could be done. The radio industry at this moment is desperately trying to find ways to make radio sexy, you know, and I think there is a bit of a crisis there. The record companies have a bit of a crisis, you know. You'd have thought two people in a crisis ought to be able to sort something out. So there is an opportunity.

The 'lack of knowledge' or lack of confidence that technological solutions could be found is again indicative of the antagonists of the entrepreneur-inventor's tale. These extracts from the two radio participants express frustration that the record industry has not wanted to work together to use technology to conceive new ways to market new products to music consumers. I now want to use two core record industry participants to illustrate in more depth the industry dichotomy of views towards technology.

Cultural industry or technology industry?

To illustrate the differences between the inventor's tale and the patron and curator tales, I contrast two passionate but quite polarized views of what record companies are, and should be. The extracts are quite long in order to illustrate the invocation of competing tales of invention, patronage and curation. Some comments are underlined to draw attention to elements of opposing polarity. The first is the independent producer/A&R:

> where record companies have got it so wrong over the last 20 years is they only exist because of technology. That's my other mantra. It only exists because Edison invented sound. It only existed because then artists put their recorded performances onto that. Then it became an art form. Things were developed for that. You know, CBS invented the long playing record. You had that art form. RCA invented the 7 inch single. Then Sony and Philips invented the CD. It's always been driven by technology and today it's driven by technology in as much as Apple invented Logic[6] and you've got Pro Tools so you can record brilliantly on here and you've got the internet which means that music can be disseminated brilliantly. But record companies are no longer in control of those two bits. They kind of have their eye off the ball. Really they should...EMI should not have got rid of its central laboratories of CRL[7] it used to be called. And they should have invented the great way of recording for everybody. And the internet side of things, they should have been, 'yeah hey, great, we can do all this'. They're insane to have got out of that because that's the driver because if we assume that recorded music is a separate creative item which it is, but it's dependent solely on technology, not actually on the artists. The record companies have got so big they think it's dependent on their personnel and on the artists. Really it's about technology. So although I'm passionate about artists, clearly I am, it is actually...the record companies are nothing more than technology-sellers and that's where they've gone so wrong, so so wrong, and you'd have thought after CD they'd have got it so so right because that gave them the hugest shot in the arm. Can you imagine if they'd leapt from CD into home recording and internet? They would now be huge multinational entertainment companies. But instead they're just these things that have had this little Canute-like, head-in-the-sand like an ostrich at the side thinking it's all going to be fine, because we control the artists.

By contrast, this next extract is from the EVP/CFO of Universal Music Group, who by his own description is responsible for 'all non-creative aspects of the Group outside North America, which means most things other than marketing or A&R':

> we should have been in these [*360 degree model*] businesses anyway. And it's only taken the decline of the CD to force us into the thinking as to what is our core…our core talent is… <u>our core expertise is finding talent and bringing it into the marketplace. That's what we do.</u> Whether that marketplace is live, that marketplace is merchandising, that marketplace is directing the consumer or that marketplace is sponsors, we should be doing all of that because that is, that is our expertise. […] there's no real reason why we shouldn't try and participate in those other businesses. And the only reason that we didn't do it before is that we hadn't quite figured that out. We made too much money let's say. Because the reality is that <u>we should have been in the artist business from the start</u>, doing everything, the old Motown model, the old MCA model which was <u>a talent model</u>. You know, those businesses used to manage all aspects of the artist's career. And there's no reason why it should have migrated away to only representing certain aspects of the artist's career. Other than commercially chasing the buck, the big buck was in the CD. It was in the recorded music. That was the big money so you chase after that, and I think wrongly excluded everything else.

Researcher (JW): I find myself nodding in agreement with all that because it sounds, you know, it sounds right, I mean intuitively right that you're in the artist business and you should therefore look at all the revenue streams. But there's another argument that goes, actually the roots of the record business obviously go back to technology and you know, it did so well for so many decades because it pioneered and developed and controlled technology and then in the, well late 80s, 90s, you know, where several of the big companies lost their connection with technology: PolyGram with Philips, and Thorn-EMI etc, Matsushita, and Sony, you know, an industry that had grown and survived on being pioneers and controlling technology. And now that that link has been broken, it's floundering. So <u>maybe it wasn't always about the artist</u> but has recognized that that is the way forward, but it's only recognized it forward because things have been so grim because of the breaking with technology? That's another argument.

EVP/CFO Universal Music: I disagree. <u>I don't think the record business has got anything to do with technology.</u> [...] I think that history is that the record business has been owned by technology companies which is a different way of putting it. The music business is just little. It's simple. It's about finding talent, bringing the talent to the marketplace. The technology aspect you refer to is the migration from vinyl to tape to the CD which was probably such a paradigm shift at that point in time, to digital, whatever that may mean, which we need to talk about at some point. But at the core of the business, <u>the record company itself should be completely technologically agnostic, because that's not what it does. It basically finds the talent.</u> But the history, you're right, the history is...

Researcher (JW): But when do you think that history changed?

EVP/CFO Universal Music: It didn't. The record business has been the record business since the start. It has been owned [...] So EMI a scientific company before, through to Philips, through to Sony, beyond... Even still I think there's an element in Japanese thinking which is that a hardware company needs to own the software company, which is also something that even in Sony's past, if you look through, has been a kind of unmitigated disaster for them, because the two businesses are so diametrically opposite. <u>And engineering, a technical background, is very different to a creative background</u> so it's not a marriage that's made in heaven at all. [...] We just need someone else to come around, but history is, <u>we've never been the inventor</u>, and someone has come up with an invention which is a consumer driven invention of which we benefit from that. And I am just very hopeful that that will happen, be it Nokia... The reality is it's probably going to be someone that we've never thought of.

These two participants reveal profoundly differing views about what a record company has been, is and should be. The polarity of the views has equally profound strategic implications and distinguishes the patron and curator tales of the Universal EVP/CFO from the entrepreneur-inventor's tale used by the independent producer/A&R. The latter asserts that the record industry's right to participate in the cultural value chain was, and still should be, only because of its mastery of technology. Take that away, and the record industry has no contribution to make. It deludes itself, he argues, if it thinks it has a natural talent for nurturing the creative process and for cultural filtration which then gives it a right to control artists.

For the first decade or so following the invention of recorded sound, the industry was described as the talking-machine industry and the commercial focus was on dictating machines. Even when it was later recognized that music would be the more influential driver of sales of gramophone players, the organizational focus of the first five or six decades of the recorded music business was as much influenced by rapid technological developments, both in sound recording and in playback machines, as by promoting musical talent, as the blurb for Gelatt's (1955) book indicates:

> This story of continuous invention, intense competition, and bitter rivalry does not neglect the celebrated artistes who [...] are as much part of gramophone history as the men who perfected the reproduction of their music.

Whilst the promotional opportunities for artists were recognized as early as 1902 with Enrique Caruso, the commercial contribution of artists and music to the evolution of the recording industry is, according to Gelatt's account, better characterized as brand and product endorsement from *established* artists, rather than as discovery and patronage of *new* talent. Even when the huge commercial opportunities were realized by the A&R and distribution model from the 1960s onwards, much of the investment was still from corporate Groups whose core strategic interest was technology-based. This was the case up until the end of the 1990s, but by 2000 Sony was the only electronics company to maintain a recorded music company as part of its strategy.

A further dimension of technology's relationship to music is introduced by another participant:

> If you think from punk to 'Dare' by Human League which I think had 5,000 edits, all done on the then most modern machine, you know, that's only a period of about 10 years, and I think those things make huge differences in the music we listen to whether we're aware of them or not. And I think there will probably be.... Sampling was probably the last great innovation in pop music and I think we're probably due another one in the not too distant future. And I think it will probably be technology-based.

> (Artist manager)

This view is that technology does not merely serve to reproduce faithfully a technology-free original, but that technology is as much part of music as the traditional constructs of instruments and vocals. The view counters the previously mentioned view of Katz (2004) that 'a discourse of realism has for more than a century reinforced the idea of recorded sound as the mirror of sonic reality, while at the same time obscuring the true impact of the technology' (p. 1). He sees the true impact as being in the aesthetics of composition and instrumentation itself, which he illustrates with many examples, from the adoption of violin vibrato in the early part of the 20th century, to synthesizers, sampling, turntablism and scratching. In this view the recorded music business and technology are inseparable.

The purpose of drawing attention to these ontological questions of whether recorded music is just a representation of some other reality, or is the 'real' end-product itself, is to illustrate how different interpretations of what music 'is' lead to different strategic positions. People telling the patron's tale and the discerning-curator's tale simply cannot see a close connection between their strategic role and that of the new media pioneers. It is the nurturing of the artist and the cultural filtration process which are paramount. These protagonists will remain *technologically agnostic* and, in flat defiance of their corporate ancestry, cannot imagine themselves as pioneers of technology.

Those with an entrepreneur-inventor bias assume that, whether through corporate acquisition, collaboration or single-minded coercion, technology can bend to organizational strategy through the process of product conception or re-conception. Had any of the major record companies been led by entrepreneur-inventors, it is possible that the music industry could have been a huge winner, financially speaking, in the new millennium. By contrast, invoking a technologist-inventor tale, in the third person, means viewing the pioneering part of technology as something outside the control and the responsibility of strategic actors within the music business, and therefore someone else's domain: *we just need someone else to come around, we've never been the inventor. The reality is it's probably going to be someone that we've never thought of.* Ironically, this same participant illustrates a more positive view of strategic agency with reference to another media industry:

Who would have thought that 10 years later, you know, there would be whatever is, 15 million households, 17 million households paying north of £40 a month to watch TV and to watch sport on TV. And you know, it's extraordinary even looking back on it but that journey that they embarked on, who would have thought that they would be able to have got set top boxes, you know, set top devices out to all of these...put aerials on everyone's bloody houses and get engineers round and help desks and PIN numbers and then start with movies but then realize it wasn't movies, it was really sport that people wanted, paying huge amounts of money to secure the football rights. And look where they are today, completely and utterly changed the whole face of TV.

(EVP/CFO Universal Music)

Universal Music Group is the clear leader in the global recorded music business and has a consistent market share in excess of 25%, and will exceed 30% in many markets following the acquisition of EMI in 2012. Along with two other companies, they collectively represent more than two-thirds of the market, and they are represented by a highly active and vocal trade body (the BPI in the UK, part of the IFPI globally). Yet despite recognizing the impressive single-minded vision and agency of BSkyB in transforming the UK TV industry, the same participant is much less confident about being able to influence the digital music environment:

you've got to figure out what is within your power to do, and what is not within your power to do, and if you think that right now, if your strategy is to grow digital, my view would be that that is not something that the record industry can, is within their power, so it shouldn't be part of their strategy.

There is a resignation amongst record company participants which assumes that they are relatively powerless to influence developments in digital technology, as illustrated by the following comment from the head of digital at a major record company (note that reference to his corporate employer is in the third person):

So they [*the record company*] go the same way. So they end up doing the same thing. To be honest I'm really not claiming to be a genius because that's all that would occur to me is that you stream it for free and you ad support it, but it's not terribly radical. [. . .] And that acceptance of reality as it is, rather than how you would like it to be, is the important bit. And I don't think there's ever been a great incentive for a company like [*my employer*] to perceive it in that way . . .

(Head of Digital, major record company)

For someone who claims to have worked at the leading edge of new media, this seems a surprisingly defeatist position from which he

Music on mobile phones – the great missed opportunity

To conclude this exploration of strategic narratives of technology, and the competing bids for role of central protagonist, I want to return to the relationship with the mobile phone companies. Several of the participants saw this as an area of optimism for the recorded music industry, and their words highlight how alternative constructions of product and technology define both the possibilities and the constraints for collaboration between the two industries. In Chapter 3, we saw the participant views on opportunities offered by integrated mobile phone services, and concluded that they face complex negotiations as the various stakeholders fight for value and power. At the heart of the struggle is that each party values its own contribution higher than the others, and wants the direct relationship with the end consumer:

> Now one thing that digital will do is that it will give us an ability to deal directly with the consumer that we've never dealt with before. And therefore let's say technology is our friend in that regard rather than our foe
>
> (EVP/CFO Universal)

The strategy head at Orange takes a different view of who should be in charge of the customer relationship:

> So the people who have the ability to link it up should be the network providers because they have the intelligence of the network and they're looking across the platform. But there's confusion between the likes of Apple, Nokia, and to a certain extent, the existing music industry trying desperately hard to market straight to the consumer.

This participant sees his own company, a network provider for mobile phone and internet services, as the logical interface with the consumer. Hardware companies (Apple and Nokia) and content companies (the record companies) are not well placed to fulfil this consumer-facing role, and by attempting to do so they are upsetting the equilibrium and confusing the consumer. The strategy manager from Nokia sees it differently:

> my preference when I was looking at this a couple of years ago was to have a company like Nokia act as a revenue collector and we would charge £3 a month and then the consumers would get unlimited music and be able to share it and all the rest of it [...] both of you

distances himself, and may explain why he left the comp
months after making the comment. The collective passivi
recorded music industry with regard to technology is a comm

> I'd have liked to have foreseen the timing and importan
> emergence of iTunes and had an opportunity to think thrc
> the relationship therefore the record companies had wit
> ahead of iTunes deciding for us all.

> (New media

The Apple relationship is a good example of the perceived im
record companies with regard to technology, though there is
complication of being accused of anti-competitive practices:

> **Researcher (JW):** maybe it [*change*] could have been accele
> know, collectively with the industry, taking a different a
> DRM perhaps but ...

> **Former CEO, EMI:** Yeah, but there's something which pe
> forget, especially the media, that there are anti-trust laws
> make any move extremely, extremely difficult. And re
> you're talking about seismic changes like this, you would
> able to make it as an industry and not as an isolated con
> it's even true for Universal. So we could not ... a lot of
> we would have wanted to make, we immediately had S
> the anti-trust authorities, both in Europe and the US on o
> that's why I think it's a bit unfair when people say, you
> haven't been reacting fast enough. [...] And EMI, with
> market share can't change the industry. And Steve Jobs (
> that very clearly.

This extract brings us back to the question of the extent to
entrepreneur-inventor can, through product re-conception
cion, politicking and protection, effect technology-based ch
saw in the custody section, some in the recorded music busin
advantageously caught up in a bigger format and interoperal
between Apple, Microsoft and others. The extract above impl
only way to take control of technology is to establish a domin
try position, such as that held by Apple in the market for di
players, but the participant believes that anti-trust authoritie
allow the music companies to act in concert to become so d
new media and technology.

[*record companies and network providers*] are having similar issues where you want to connect with your customers and you have sort of, dwindling in their eyes, value proposition and inability to really create new services on top of it or not being as effective at creating new services on top of it.

This proposal would have put Nokia, which at the time of the conversation was the largest mobile phone manufacturer in the world, firmly and competently in the front-line, illustrating the strategic and competitive desirability of doing whatever it takes to gain and maintain control of the consumer relationship.

The record companies fought the phone companies on the consumer-access issue, but they could imagine conceding:

> the great belief at one point was that this was going...one of the fundamental shifts was going to be that the record companies themselves would have a direct relationship with the consumers. And we thought we were going to be the gateway to the consumer, and therefore we wouldn't allow the telecoms to get between us and our consumers. And so the big sticking points in negotiations became access to the consumer database and the telecoms would say, 'no bloody way. We're not giving you that.' And we'd say, 'well we're not giving you our music then'. You know, all of this standoff. And I think another piece of pride swallowing will have to be that we have to recognize that actually probably that fundamental shift hasn't happened and that there are...that effectively the record companies are still dealing with retailers. It's just that those retailers are different people. They are phone companies...
>
> (new media financier)

Universal had a similarly difficult experience with the mobile phone network service providers, as we saw earlier in the reference to Vodafone as the '800 pound gorilla bullies':

> Everyone is trying to be something that they're not. And you can't be that. So if you're a mobile company, you're not a content company and you shouldn't be. [...] what happens is if you're a Tel Co and you want to take 50% of margin where you don't incur 50% of the cost it doesn't work. So if you're a distributor you should take a 10% margin and be done with it, but that's not their view of what they should be doing. Japan's quite different by the way. So in Japan the Tel Cos basically provide the environment and it's the most successful digital

content business in the world and it's because the telcos behave as distributors and not as content...

(EVP/CFO Universal Music)

As with the Nokia and Orange participants, the Universal participant has equally clear and fixed views of identities, products, services and fair division of proceeds with regard to distributors and content companies. But when *everyone is trying to be something that they're not*, the narrative worlds can't hang together, and either bad things happen, or nothing happens. Though Universal had made some progress with Nokia on their 'Comes with Music' project, the participant continues to be exasperated with Vodafone. Underneath the exasperation with Vodafone's bullying stance, there is an exasperation that they are failing in what he perceives as their responsibility to innovate:

> it's [*growth in digital music revenues*] nothing to do with the record industry [...] It's to do with others. The others figure it out, we're very beholden to them. Now you can sit down and you can try to help them along the way, and you can make things available, you can give them licences, you can give them content, you can give them exclusives, you can do whatever. But until someone comes up with a product offering that's compelling to the consumer, then it ain't gonna work. We spent years trying to help Vodafone figure all this stuff out.

He explained that Universal was working on a collaboration with SonyBMG (the second biggest record company) to provide a subscription service through an alliance of all content-owners, but:

> it's very problematic...I just fear that that's going to fail.

Once again, the feeling is that even a collaboration of the most powerful companies in the recorded music industry cannot master technology alone, and that they are 'beholden' to others to reconceive the consumer offering.

I will leave the last participant word in this section to the strategy head at Orange, who seems to me to best sum up the state of play. It dates from 2008, but is still relevant in 2013:

> I think at the moment there are too many people bouncing around the place and, yeah, which is leading to a lot of confusing messages in the market, which is leading to over-promising and generally under-delivering and what is deemed now convergence is what

I would loosely call an aggressive bundle. It's not at a user level yet and we've discussed, you know, mobile companies coming at it with a mobile phone approach, handset companies coming at it from a handset side and we've got guys from the existing business looking to deal with these people, but it's not joined up. And I think there's still an opportunity to have a really, you know, people to have…partnerships are difficult things but if you are absolutely clear what you're trying to achieve by that partnership, if you say, Sony BMG and you say, right, we have this type of content which is really appealing to this type of person and you as Orange say, we've got this type of handset range and this type of tariff and this type of marketing campaign that you, Sony, can leverage with us to this, exactly the same type of person, then, you know, you're complementing each other and the business model is you know, transparent and agreed. And I think you know, that sort of proposition where both of you have clear aims and focus could work very well.

(Head of strategy, Orange)

There are many examples in the texts of participants expressing their desire for greater collaboration. Equally, there is an intellectual vision that, with the combined skills and assets of record companies, network providers and handset manufacturers, compelling new products could be marketed to consumers. However, this 'happily ever after' ending may be a commonly desired conclusion to what look like superficially similar inventor's tales, but there is a good deal of competition for the role of central protagonist or hero. There was certainly no happy ending for Nokia. Its lack of a compelling music service was symptomatic of the obstacles to strategic collaboration between content and technology, and is typical of the previous and subsequent failed strategic initiatives between record and phone companies which never managed to move beyond simple licensing deals.

I have used the narrative frame of tales in this section to outline different identities and views of the world which give rise to different attitudes to strategy and change, and to shed some light on the question of why so many solutions to the industry's problems have failed to be commercially executed, despite having been conceived many years ago. In the next chapter I conclude Part III of the book with critical reflections on issues of power and ideology, and consider the bigger political issues at stake in the storytelling contest.

11
Power and Ideology

What is the music industry?

Given that so much discourse presumes the real tangible existence of something called 'the music industry', it is useful to conclude Part III of the book with a macro-level cognitive question of industry identity. Williamson and Cloonan (2007) draw attention to the political dangers of defining music as a single industry when it includes so much diversity, being concerned in particular that the recorded music industry is too often treated synonymously with the whole music industry. This view obscures the fact that live music promotion and publishing, for example, have quite different characteristics from recorded music, not least in that they have had diverging economic trajectories in the past 12 years. The authors explore the way that four different groups use industry definitions, either unconsciously or deliberately, to serve their own purpose, namely: trade and representative bodies; media; government; and academics. From this, they report various reasons for refraining from the singular usage of the term 'music industry'. Three of these are of particular interest here, namely:

- inequality
- conflict
- policy

'Inequality' refers to the fear that seeing music as just one industry over-privileges not only the dominant logic of the major record company model, but also a particular structure based on multi-national operations. The interests of individuals and of small independent businesses are therefore neglected. 'Conflict' refers to the misplaced assumption that all players within the industry have common interests. The authors point to the many conflicts in the industries, such as those between artist and label, promoters and venue owners, and the recording industry and the music publishers. Finally, 'policy' refers to the danger of government treating the music industries as having homogenous interests: 'the DCMS[1] cannot help an industry until they know what it is' (p. 318). They conclude that 'the nomenclature serves only to reinforce ideas of a single music industry dominated by the large record companies, and assists industry organizations in attempting to impose their worldview'. Williamson and Cloonan's reference to a dominant, or even an imposed, worldview echoes the comments of the UK government minister:

> People have got very entrenched views about that. I mean, these are theologies for some people.

He is actually describing all sides of the debate on music copyright protection, including those who seek liberalization of copyright protection, and not just those who are lobbying to extend it. Theologies may seem like a strong term, but there are certainly some powerful belief systems in evidence throughout the conversations. The tales of patrons, curators and inventors are not isolated stories, but reproductions of sense-making mechanisms which confirm or challenge whole value systems about the way society should operate, and in this regard they go well beyond the music industries, even as broadly defined by Williamson and Cloonan.

My analysis thus far has tried to locate discursive practices within the wider societal context of grand or mega discourses. The purpose of using the seven tales is to illustrate both the constructive and restrictive power of the prevailing discursive positioning of those with a stake in the economic future of music. Burr (1995) describes such discursive positioning:

> Once we take up a position within a discourse (and some of these positions entail a long-term occupation by the person, like gender or fatherhood), we then inevitably come to experience the world and ourselves from the vantage point of that perspective. Once we take on a subject position in discourse, we have available to us a particular, limited set of concepts, images, metaphors, ways of speaking, self-narratives, and so that we take on as our own. (p. 145)

The tales create identities which can be as strong as Burr's examples of gender and fatherhood. These identities make available ways of seeing and ways of acting which can become so well established that challenging them appears inconceivable. Discursive positioning is closely linked to institutional practices in a reciprocal process through which organizations are sustained by discourses, which in turn then validates them. Whether this is described as a virtuous or vicious circle depends on the ideological position of the interpreter.

The discourse which sustained and reproduced the recording industry for the first hundred years of its existence was born of an entrepreneurial-inventor's position, which used the medium of recording and playback technology to construct recorded music as a cultural product which is independent of its source. It also introduced music consumption as a private experience, rather than a public one. This was a fundamental shift from music's previous cultural and social functions, which brought people physically together through live performance. The proliferation in the channels of dissemination brought about by this new industry expanded both the number and the scope of the roles of patron, businessman and curator. In these tales, a primary objective is to overcome the economic challenge of music being a non-rivalrous[2] good and to protect and enhance its premium value. The statutory protection of intellectual property through culturally embedded copyright provisions has for the most part served this objective well until now, but there have been earlier substantial threats to the sustainability of economic value in the recorded music business. The struggles arising from these threats are worthy of consideration.

One of the most illuminating genealogical perspectives comes from radio, in particular its relationship with the recording industry, which has oscillated between competitor and collaborator and has been much commented on (Gelatt 1955; Huygens et al. 2001; Katz 2004; Dowd 2006). During the American depression, the music which was freely available on radio was thought to be threatening to make obsolete the concept of recorded music as a product to be purchased and owned. As we saw in Chapter 1, between 1927 and 1932 the number of discs sold fell from 104 million to 6 million. The big three record companies concluded that record firms were suffering from radio broadcasts of recorded music. They opposed broadcasting of their products rather than encouraging radio stations to air live broadcasts of their performers in order to promote sales. In 1933 they even started to inscribe prohibitions on their discs[3] (Dowd 2006). Government policy endorsed this move, as did the parent broadcast networks of the radio stations, which championed live programming as preferable to recorded music, but by the early 1940s these restrictions had been removed as they were practically unenforceable. Taking a radically different approach, a small newly established company (Capitol Records, later acquired by EMI) believed that radio broadcasts would stimulate sales and began to send free promotional discs to radio stations. By 1945, the practice was so institutionalized that the industry measure for success (the Billboard chart) incorporated radio airplay figures.

This historical reference is relevant to the process of critical reflection. The impact of the internet is regularly described in terms which are reminiscent of the fears felt by the recording industry towards radio in the pre-war years. Alternative discourses which conceive of radio broadcasting in promotional terms similarly resonate with the Wiki world arguments that the internet is a communication mechanism of extraordinary power and reach. Just as Capitol Records saw the opportunity to reconceive the technology of radio, so new media entrepreneurs pursue the value-creating opportunities afforded by the internet and mobile phones.

The economic value of Capitol's strategy became particularly apparent in the decade between 1950 and 1960, when the majority of sales value had moved from single songs to long-playing (LP) records which were promoted by the massive exposure of those single songs. The technology which allowed longer recording time, combined with the more favourable economics of selling bundles of songs (albums), came to define not only the product but also the artistic process and the production and marketing schedule (Keightley 2004). Record companies became focused on the longer-term developmental potential of artists, with contracts binding artists to multiple album deals extending over many years. In what Miege (1989) called 'the dialectic of the hit and the catalogue', increased value lay not only in selling bundles of songs on the back of one hit, but also in developing a more enduring popularity of an artist's back catalogue. As we saw in the participant comments, the specific technological-historical origins of this dominant product conception tend to be overlooked or forgotten, and the cultural status of the album format and the concept of artist career are seen as ontologically more absolute and timeless than a historical view would support. This is especially so amongst the baby boomer stakeholders who grew up with this product conception.

The historically evolving product conception is also identified by Huygens et al. (2001), who trace the strategic development of the recording industry from 1877 to 1997. From the perspective of organizational co-evolution, they loosely identify five eras differentiated by the shifting focus of inter-organizational competition:

- 1877–1914: competition for hardware (playback equipment and disc format)
- 1914–1930s: competition for software (existing & established artists to endorse hardware)
- 1930s–1950s: competition for markets (new genres, artist development and 'star' system)
- 1960s–1980s: competition for labels (new artist/genre specialists feeding global distribution)
- 1980s–1990s: competition for catalogues (acquisitions for scale economies and CD boom)

This evolution, and in particular its pursuit of scale economies for physical distribution, has resulted in a declining number of companies controlling an ever larger share of the market.[4] From the 1950s, the product conception was stable and, I would argue, hegemonic. Since 1999, the economic consequences of the digital unbundling of the album format, together with the decline in the premium value attributed to the physical product, have reversed the scale economies on which the major record companies' infrastructures are built, and are thus financially very damaging. It is therefore unsurprising that traditional stakeholders are reluctant to give up the dominance of their product conception and the associated statutory rights which underpin their identity as independent cultural products which can be owned and controlled for long periods of time.

Discursive power and the struggle for survival

The music industry is routinely portrayed as fighting for its survival. As illustrated by the texts quoted in this research, the battle is largely being fought with discursive weapons. Alternative and competing conceptions of products and services are inextricably linked to embedded and sometimes ideological ways of seeing the world and acting within it.

The interpretive repertoires underpinning the object constructs of music, record companies, technology and the consumer can now be seen to draw from several grand/mega discourses. The same is true of the protagonist and antagonist identities of the seven tales, and of my constructions of Tin Pan world and Wiki world. For example, a discourse of technological determinism favours a view of social evolution which diminishes the influence of human political agency and only acknowledges the contribution of the inventor as a catalyst for a process which would happen sooner or later anyway. A discourse of capitalism underpins the sanctity of intellectual property rights and incentivizes large-scale patronage and curation. Neoliberalist discourse favours the entrepreneur-inventor who can shape markets with minimal government intervention or constraint. Liberalpluralism provides interpretative repertoire for the liberal-curator and disseminator in the way it highlights the damaging social consequences of narrow concentrations of economic power and control in cultural production.

Returning to Fairclough's (2001) suggestion that critical discourse-based studies should be focused on a social problem, that problem is, loosely, the custodial tensions illustrated in Chapter 4 and which are also illustrated in the tensions between Tin Pan world and Wiki world. Fairclough's recommended approach is then to identify obstacles to the problem being tackled and to consider whether the social order 'needs' the problem, by asking such questions as: who has an interest in it *not* being resolved? Some of the new and growing stakeholders in music have an interest in discursively devaluing the record companies' contribution to the value chain, i.e. they have an interest in the non-resolution of those issues constructed by the music industry as problems. The various participant interpretations of the public discourse promoted by Apple's Steve Jobs on subjects such as digital music pricing, DRM and subscription models provide good examples of such discursive dynamics. The arguments used for and against the concept of disclosure and action by the internet service providers to prevent the illegal copyright-infringing activities of their subscribers, in opposition to the 'mere conduit' and 'safe-harbour' protections of EU and US law, are similarly revealing with regard to who has an interest in the problem not being solved. But the mother of all discursive battles in this domain is the one which goes to the heart of the matter and concerns the validity of the concept of intellectual property.

Practically speaking, intellectual property is an economic issue which, could one turn back the clock half a millennium, one might imagine could be solved rationally or empirically, without emotion. However, the centuries since Gutenberg invented the printing press have witnessed the establishment and reconstruction of intellectual property in ways which become so bound to identity, wealth, social status and political influence that one cannot address the question without reflecting on philosophical aspects of the phenomena of intellectual property. The concluding section of the book, Part IV, attempts to address the discursive properties of intellectual property under the heading: The Pirate's Tale: The Reform of Copyright, and the Future.

Part IV

The Pirate's Tale: The Reform of Copyright and the Future

Part I of the book was heavily influenced by my own experiences and narrative interpretation of music industry history. Parts II and III were centred on the discourse dynamics of a variety of stakeholders, in order to gain insight into the identity-bound constraints and obstacles to change which face the industry. In this final section of the book, I focus on what I believe to be the most important, or at least the most interesting question which will affect the future of the music industry, and indeed the whole of the cultural economy: the reform of copyright. In doing so, I endeavour to link the analysis and insights of previous chapters to copyright discourse in the broader public sphere. An evaluation of the 'fitness-for-purpose' of copyright in the digital age involves a consideration of the historical evolution of the law and its infringers, in the context of the prevailing economics, politics and philosophies of the last three centuries. This part of the book must therefore go well beyond the music industry, and look to earlier technological transformations in book publishing, and to the contested history of the legitimacy of authorship.

Chapter 12 picks up on a traditional counter-narrative which was notably absent from the participant conversations of Part III, namely the Pirate's Tale. It charts the historical origins of the construction of the pirate and how it has been transformed in the last 15 years to suit alternative ideologies and political agendas. It explores the popularity and moral ambiguity of pirate narratives and how they play an important role in questioning the relationship between the individual and the state in matters of creativity, knowledge-sharing and trade.

Chapter 13 introduces the challenge to copyright from a newer counter-discourse of cultural conservation and of the public domain. I suggest that this discourse lacks the overarching framework and emotional connection of a master-narrative. It is more defined by its various antagonists than by a clearly drawn protagonist and that, despite its notable impact, it is consequently falling short of the worthy aims of its proponents. It concludes that unless there is a radical shift in public perception towards the alleged injustice, inefficiency or corruption of intellectual property law, then originators and their corporate patrons and protective-curators will continue to tell more successful tales than the liberal-curators and disseminators, the latter remaining characterized as troublemakers, thieves or pirates.

Chapter 14 builds on the counter-discourse of chapters 12 and 13 and presents a brief history of copyright law. In an attempt to gain some perspective on whether the unchanging institution of copyright is really a problem or not in the digital age, it is useful to take a much longer historical view of human knowledge creating processes. I reflect upon whether the post-Gutenberg construct of the literary mind represents social progress which must be preserved, or whether it has become obsolete in the digital age with new forms of knowledge-transfer and 'orality'. The chapter concludes by considering the likelihood of copyright being reformed in the second decade of the new millennium.

Chapter 15 concludes with some reflections about the future of the music industry and of cultural production in general.

12
Pirates, Property and Privatization

> No black flags with skull and crossbones, no cutlasses, cannons,
> or daggers identify today's pirates. You can't see them coming;
> there's no warning shot across your bow. Yet rest assured the
> pirates are out there. [ ] The pirate's credo is still the same –
> why pay for it when it's so easy to steal
>
> (RIAA website MusicUnited.org, cited by
> Reyman 2010, p. 63)

When I joined the music industry in 1992, the unauthorized copying
of recorded music was not top of the list of economic threats to the
corporation. Bootlegging, which is a term used for the exploitation of
unauthorized recording of live music concerts, was at least as big a concern, especially in the US. As for the unauthorized copying of *authorized*
recordings, the campaign which had been launched in the 1980s under
the Jolly Roger logo of a cassette and crossbones,[1] and which declared
that 'Home Taping is Killing Music', had run its course. It was parodied,
and at least partially discredited by the sustained healthy growth of the
music industry in developed markets.

As for commercial 'pirates' (as opposed to the home-taping domestic
ones), the industry was well aware of their unauthorized cassette and CD
manufacturing activities, which had been going on since the invention
of the recording formats, especially in Latin America, Eastern Europe
and Asia. The scale of such activities was relatively measurable. Fighting piracy was a priority of the industry trade bodies (RIAA, BPI and
especially the IFPI) and their anti-piracy initiatives and reports were certainly something to be monitored. However, the overall feeling at that
time was that piracy was manageable, especially when pirate copies were
often physically distinguishable and inferior to the authorized versions.

It thrived mostly in foreign markets which were still underinvested by the major record companies. Ironically piracy served, to some extent, as an informal and unacknowledged source of market-research and market-making prior to increased investment in some territories. Once commercially organized piracy had grown to a certain level, it was indicative that record companies could justify asserting their rights and investing more heavily in that territory, always presuming that the local government had an appetite for enforcing its copyright laws. It was also indicative of which artists and genres were most in-demand in those markets. But to put all of this in the perspective of the time, even if some territories had worryingly high physical piracy rates, it was believed that, at a global level, piracy represented less than 10% of the total market. As long as it was kept below that level, it would not substantially inhibit the continued growth of the authorized market.

By 2000, the industry view of piracy had changed dramatically. Home-taping, and even CD ripping and burning were to some extent measurable by sales of blank units, but peer-to-peer file-sharing played on the worst fears of music executives: not only could they not control the dissemination of music, they also had no idea how to measure or monitor what music consumers were actually doing. All they knew was that millions of people were accessing file-sharing sites, and that hundreds of millions of digital music files were being shared. In the worst case, if one were to judge music consumption simplistically by the number and source of music files on iPods and on computer hard drives, one might have imagined that the market had inverted, and was now 90% 'pirate' and 10% authorized. Such a change in the perception of the threat of unauthorized copying called for a drastic response, in both actions and words.

There have been several strands to the record companies' digital anti-piracy strategies including legal actions, technical measures (encryption, DRM, spoofing, surveillance), public education, government lobbying, and the development or promotion of authorized digital services. In the short term, the focus was on trying to constrain or disable, rather than enable, new forms of consumption. An aggressive litigation strategy, especially through the RIAA, is well documented and decried in the blogosphere. Starting in 2000 and continuing to the present, lawsuits have been brought against a sequence of file-sharing sites: Napster, Grokster, Kazaa, Limewire, eDonkey, Pirate Bay, Megaupload, to name only a few. These actions have, in isolation, been mostly successful, but as each one was forced to close, others sprang up to satisfy the huge demand of the burgeoning file-sharing community. Legal actions against deviously clever file-sharing services were very unpopular with active file-sharers, and regarded by new-age technologists and web-savants as futile given that the infrastructure of the internet does not recognize territorial customs and laws. Nevertheless, the traditional industrial logic was clear enough: seek the protection of the law against alleged infringements of copyright and shut the services down.

In contrast to the understandable industrial reaction of pursuing identifiable file-sharing services, a parallel strategy drew much less, if any, public support. This strategically questionable and alienating development was the pursuit of .individual users on a grand scale. Exploiting provisions of the 1998 Digital Millennium Copyright Act (DMCA), subpoenas were issued in the US to internet service providers (ISPs) requiring them to reveal the names and identities of alleged infringers. Huge numbers of individual lawsuits followed, and between 2003 and 2006, the RIAA had brought lawsuits against more than 17,000 individuals. Public opinion soon turned against the industry when unwitting or naïve users such as young children and grandmothers were caught in the dragnet approach.

Whilst legal action against consumers, persistent up-loaders and file-sharing services has continued to this day, the music industry has moderated the aggression. Another focus of its attention was to lobby for statutory solutions which make it easier for content-owners to get ISPs to cooperate in giving fair warnings to infringing users which, if ignored ('3 strikes'), could lead to the suspension of the users' internet service altogether. Such a provision was one of the most notable elements of the UK's Digital Economy Act (2010) but has proved slow and contentious to implement, and may well be abandoned. Consistent with its reputation as a nation with high standards and rules of cultural and intellectual property protection, the French government was the pioneer of the three strikes concept in 2009 with the previously mentioned law known as Hadopi. However, even in France the law provoked anger and was revoked in July 2013, amidst questions of its effectiveness and arguments that the punitive penalty of cutting internet access was draconian and disproportionate to the offence. The idea that internet access has become accepted as a basic human right was evidenced in 2012 when a UK Court of Appeal reversed an order which was preventing a convicted child sex-offender from using the internet. The judge remarked that it was 'entirely unreasonable to ban anybody from accessing the internet in their home'.[2]

There is already much written on the topic of record company reaction to file-sharing, and I do not intend to go into detail on the legal developments. Of more relevance to the theme of this book are the moral question of copying and the narrative of piracy which has been adopted in these corporate digital anti-piracy strategies, especially in the 'education' of the public about copyright infringement. The US has been more active than the UK in this regard, and the film industry has been much more visible in its campaigns than the music industry. Perhaps the best-known example is the short video (49 seconds in length) produced by the Motion Pictures Association of America in 2004 and included for several years thereafter as part of the anti-piracy warnings on DVDs and amongst trailers in cinemas. Against a music soundtrack which includes a police siren, a girl sits in a bedroom downloading a movie, interspersed with footage of various types of street theft with the captions:

> You wouldn't steal a car; you wouldn't steal a handbag, you wouldn't steal a television; you wouldn't steal a movie. Downloading pirated films is stealing. Stealing is against the law. Piracy. It's a crime.

Consistent with the forced viewing of the ubiquitous copyright warning text which was incorporated into DVDs, this video could not be skipped, meaning that legitimate purchasers of the video were forced to view it each time they viewed the movie. It was heavily criticized and, like the old 'Home-Taping is Killing Music' campaign, it has been much parodied. In an ironic twist, the composer of the music commissioned for the video brought a legal action based on the fact that it was only intended for use at a film festival, and its wider copying and exploitation within millions of DVDs was therefore unauthorized.[3] Although he was ultimately successful in his legal action, it was a long and messy process. This is a good example of the problem with copyright law: it is complex to understand and to administer and, more importantly, the degree of 'wrongness' of any given instance of copyright infringement is difficult to nail down, and therefore to legislate for, and to prosecute. For these reasons, it is well worth examining the societal role of the pirate, historically and symbolically, in more depth.

The pirate as cultural icon

It is clear from all of this that the underlying ethical question of copying is as opaque as the laws which try to encompass it. The word pirate is similarly ethically opaque, consistent with the diversity of its etymology. There is a cachet about pirates which is nowhere better illustrated in recent times than in the character of Captain Jack Sparrow, richly developed by the actor Johnny Depp in the spectacularly successful Walt

Disney film franchise, *The Pirates of the Caribbean*. As a 21st-century cultural icon, Sparrow has received some impressive tributes, according to his long Wikipedia entry: 'the only iconic film character of the 2000s decade' (Emmanuel Levy); 'the only element of the films that will remain timeless' (Todd Gilchrist); 'definitely one of the most dazzling characters of the decade' (Entertainment Weekly); 'precious few film characters have epitomized what makes the outlaw such a romantic figure [as he] lives for himself and the freedom to do whatever it is that he damn well pleases' (IGN).

Those who argue that Generation Y, or Millennials, are demographically characterized by narcissism and a strong sense of entitlement[4] would no doubt see Sparrow's enormous popularity, despite his selfishness and trickery, as consistent with their propositions. Most unauthorized downloading happens despite people knowing that what they are doing may be illegal,[5] and various techniques are employed by 'pirates' to justify their behaviour. Ingram and Hinduja (2008) demonstrate that these techniques are consistent with Sykes and Matza's (1957) techniques of neutralization and theory of delinquency and include: denial of responsibility, denial of injury, denial of the existence of a victim, a tendency to condemn the condemner (e.g. the record companies for their greed, control-freakery or incompetence), and an appeal to a higher loyalty or ideology. However we theorize it, we should not be surprised that a combination of social factors have all come together to invite disruption of the old model. These factors include: desirable social products marketed globally to many people who cannot afford them, perceptions of the over-pricing of authorized products, the ease of availability of free copies, mistrust of big media corporations and allegations of their hypocrisy, and the aggressive application of laws which are confusing and may no longer be fit for purpose in a digital age. A gap between the rules of law and the new social norms shaped by technology has arisen, and in this gap an alternative worldview is fostered, where peer-group values and individual conscience are higher authorities than an anti-piracy warning on a DVD. In this context it may be an ironic 'own-goal' that Disney, as one of the most robust defenders of its copyrights and a vocal lobbyist for stronger copyright protection, has at the same time romanticized the pirate so effectively for the downloading generation.

The origins of the intellectual property pirate

There have been claims (Gantz 2005) that the first intellectual property pirate was Saint Columba, the 6th-century Irish monk who made copies of his Abbot's psalter without permission. The dispute was about political power and the control of the spread of Christianity in a pagan society, thus the Abbot looked to the local King Diarmait for resolution. With the line 'to every cow its calf, to every book, its copy', he ruled in favour of the Abbot and against Columba. Diarmait's analogy of birth and lineage does raise an interesting question about the relationship of copies to their originating 'master', but as there is no evidence of the use of the word pirate or piracy, this incident may be better characterized as the first account of an institutionally remedied copyright infringement, rather than an account of piracy.

The use of the word piracy in the context of theft of intellectual property does not emerge until the 17th century where it becomes a useful construct for those vying for control over publishing. Importantly, it pre-dates the construct of authorial or intellectual property. As recounted by Adrian Johns (2009, pp. 29–40), by the middle of the 17th century, the legacy of the English civil war was a traumatized society and an intellectual anarchy where the old legal and administrative structures and customs of the book trade had broken down. The popular press was 'viciously partisan, violently sectarian, ruthlessly plagiaristic, and often wildly credulous' (p. 30). In this context, the pirate became a necessary *invented antagonist* in the establishment of publishing integrity and rigour, and more generally in the fight to define civilized society. Whilst there was broad consensus about the print pirate's destructive influence over civilized society, there was a struggle to define and to claim the singular role of the worthy *protagonist* in this narrative.

The incumbent candidate was the community represented by Stationers' Company, a Crown sanctioned institution which for almost 200 years had enjoyed a significant influence and control over the new technology of printing, including censorship. This institution had lost its exclusive control in the civil war years. Milton's *Areopagitica* (1644) is an example of the more worthy and eloquent attempts to encourage a vibrant public sphere and to establish freedom of the press during that period. But unlicensed printing was equally seen as anarchic and destabilizing, even as a weapon.[6] The Stationers' Company's control was therefore restored by the Licensing of the Press Act in 1662, which was largely driven by the desire of the newly restored Charles II's parliament to prevent 'Abuses in printing seditious treasonable and unlicensed Bookes and Pamphlets'.[7] Despite this apparent alliance between trade and state, there were, amongst those loyal to the Crown, some who believed that printing had become a corrupt oligarchy of booksellers and should be radically reformed or eradicated. The alternative proposition, spearheaded by an old Cavalier called Richard Atkyns, was a revival of the royal patents system, whereby printing was a privilege granted by the Crown to 'gentlemen' who, he argued, were less corrupted by the mercenary interests of trade and speculation. Atkyns claimed to be the rightful heir to a patent granted by Elizabeth I for all books on the common law. During the civil war, the Stationers' Company had taken effective control of the publication of common law books through their entry in the Stationers' Hall register. The control of common law book publishing was particularly valuable in a society which was struggling to re-establish legitimacy and authority. The battle was thus a profoundly political one in which all publishing interests were at stake. Each claim rested on different concept of 'property'. For Atkyns' argument to be credible, he needed to rewrite history and argue that the art of printing belonged to the Crown. He did this by claiming (Atkyns 1664) that Henry VI had commissioned[8] the creation of a community of printers as Crown servants, prior to Caxton's introduction of printing to England, as an autonomous activity of private craftsmen. Printing therefore 'appertaineth to the Prerogative Royal and is a Flower of the Crown of England' (Atkyns cited in Luckombe 1771, p. 8) and was therefore subject to royal control through licence or patent.

According to Luckombe, Atkyns' claim was dubious. To counter Atkyns' new narrative, the Stationers invented their own concept of property: authorial property, whereby an author was said to have an absolute right to his work. By contracting with the bookseller to register it at Stationers' Hall, it would be preserved in perpetuity through policing by the bookseller. As authors were rarely the beneficiaries of the Stationers' registration process, this was a radical departure from prevailing concepts of author rights. Johns (2009) argues that:

> this may be the earliest explicit articulation of the idea of literary property – of an absolute right generated by authorship, which could serve as the cornerstone of an entire moral and economic system of print. Certainly the idea had no clear precedent behind it. It was nowhere referred to in the [Stationers'] company's own founding documents, nor in the century long record of negotiations at its court, nor in the broader legal arena. (p. 38)

Three hundred and fifty years of discursive reasoning later, we are comfortable with some form of authorial property as common sense. By contrast, in the late 17th century, the idea that any and all writers could claim natural rights of ownership and control over whatever they wrote was a radical and provocative challenge to other forms of authority, legitimacy and social order. The author as original creator, and owner of natural rights was far from being established as the logical hero-protagonist in a narrative which already had such a clear villain-antagonist. Unauthorized copying was seen as socially destabilizing and antithetical to civilization. Those who engaged in it needed to be identified by an appropriate label. 'Pirate' fitted the bill and fell into common usage relatively quickly and uncontentiously. But concepts of authorial property rights were newer, more contentious and took much longer to be adopted.

Atkyns' argument prevailed in the short term and the Stationers' charter was revoked in favour of the system of gentleman patentees, though not until after Atkyns' death in 1677. Following the removal of King James II in 1688, the new government of William and Mary allowed the Stationers' licence to lapse in 1695, and from then on a new era began, where the notion of natural rights of the author would gain currency, leading to the Statute of Anne in 1710. The Statute is widely regarded as the first copyright law, and became the influential source of further laws, both in Britain and overseas. Nevertheless, it is relevant to reflect on the conundrum that the intellectual property pirate was effectively invented *before* the 'property' he was alleged to be stealing was clearly understood, defined and accepted.

The golden age of piracy

Given the important role played by the pirate metaphor (if it still is a metaphor), in constructing and reinforcing the concept of intellectual property, it is interesting to consider the broader cultural origins of the 'original' pirate in his non-intellectual property sense. As for the etymology of the word piracy, Adrian Johns (2009) claims the word derives from a distant Indo-European root meaning a trial, attempt or experiment. He argues that its association with seafaring lawlessness became prominent through Thucydides' (c.460–c.395 BC) account of the suppression of the *peiratos* as a key element in the rise of Athens, before which piracy had been seen as honourable. After that time, pirates were routinely portrayed as 'irritants to the civilized order', and thus 'civilization was the antithesis to piracy' (Johns 2009, p. 35). For Cicero, a prominent denouncer of pirates, it was not their violence, but their unsociability that seemed to be their defining characteristic, and he therefore didn't differentiate seafaring pirates from land-based ones. He drew an important distinction by seeing pirates as situated beyond, rather than within, society. Perhaps because the sea provides a more suitable domain for remaining beyond society, the ship-based pirate has acquired a stronger cultural identity.

In what he describes as a genealogy of Captain Jack Sparrow, rather than a history of piracy, Martin Parker (2009) traces the fascinating relationship between historical and fictional representations of piracy, which he finds 'too intertwined to be disentangled' (p. 169). Though there is no 'truth' of piracy which is easily accessible, its varied representation is useful as a medium to raise questions about the relation between the individual and the state and the boundaries and the legitimacy of trade. The 'golden age' of piracy is generally regarded as Atlantic piracy in the 17th and early 18th centuries. This may explain its parallel emergence as a metaphor for unauthorized copying over the same period. It is thus worth considering what it was that made piracy so culturally rich and colourful.

Seafaring pirates of the golden age owe their existence to the colonizing and trading ambitions of European nation states. It is the territorial limits of nation states and the practical un-governability of the oceans which permit pirates an alternative existence 'beyond society'. Pirates can be found at the extreme end of a range of activities which begins with privateering, sometimes referred to as state-licensed piracy. Privateer is another word which emerges during the 17th century, and is defined as

> an armed vessel owned and crewed by private individuals, and holding a government commission, known as a letter of marque, authorizing the capture of merchant shipping belonging to an enemy nation.
>
> (Oxford English Dictionary)

Captains and crews of these vessels often went beyond their state-sanctioned commissions for their own gain and pleasure, and it is the extent to which such transgression is tolerated or punished which is of most relevance to the theme of this book. Privateering was an efficient mechanism for economic and political expansion and it also helped to minimize the costs of maintaining a navy. This usefulness meant that the transgressions of privateers were often overlooked. Queen Elizabeth I's relationship with Sir Francis Drake, whom she allegedly addressed as 'my dear pirate', is one example of how much leeway state-sponsored pirates had, and that their activities, though often illegal and violent, did not stand in the way of royal favour and career advancement.

If the golden age of piracy began with a sense of national loyalty, tales of derring-do, and a convenient blurring of rules and law enforcement, it later evolved into something portrayed as much more cruel and violent, and certainly not the feasible pursuit of a gentleman. The law became ever more aggressively enforced, leading to escalations of violence between pirates and the authorities. What drove the change was the rapid growth, in operational scale and political power, of the merchant classes and their international trade. Growth required the investment and the protection of capital, and this required the strict enforcement of law and order. By the end of the 17th century, the merchants had formed alliances with aristocrats and state bureaucrats which meant that 'there was to be no place for pirates in this new world, no place for individualist marauders on the periphery of empire' (Earle 2004, p. 146).

Being outlaws, and therefore beyond the 'protection' of the law, pirates could be hung wherever, and by whomever, they were found. The powers that sought to stamp out piracy constructed the notion of a 'universal jurisdiction' which could ride roughshod over foreign state laws and customs. This development had, and still has, a profound impact on international law. Kontorovich (2004) shows how 20th- and 21st-century advocates of a new universal jurisdiction ('NUJ') have sought to establish its legitimacy by 'invoking piracy as precedent, justification and inspiration' (p. 184). The justification of NUJ is based solely on the heinousness of the alleged conduct. War crimes, torture and terrorism tend to be the focus points, but encouraging people to think of Al-Qaida as trading pirated DVDs for flying lessons suggests that continuing to associate piracy with horrific violence is a convenient rhetorical resource for those industries which are most vulnerable to piracy (David and Kirkhope 2005, p. 90). Kantorovich argues that, in using piracy as its precedent, courts, scholars and political proponents of NUJ have uncritically and selectively accepted portrayals of piracy as heinous crimes, when in fact historical analysis would suggest that 'piracy was not regarded as particularly heinous'. Nevertheless, an emphasis on the vicious and violent aspects of pirates continues to support the metaphorical power of piracy in claims about the universality of international law and the protection of international trade.

Fictional heroes or villains

In his introduction to Captain Charles Johnson's *The General History of the Robberies and Murders of the Most Notorious Pyrates, and Also Their*

Policies, Discipline and Government (Johnson 1998), David Cordingley states that the modern conception of pirates comes from this highly influential 1724 book, of which there are 70 editions (Burl 2006). Cordingley claims that 'the majority of the facts have been proved to be accurate' (intro., p. IX), though others have argued that it is 'more fiction and disguised social criticism than serious history' (Parker 2009, p. 168). Either way, the book has had an unquestionably dominant influence on the popular images of the golden age pirate. In the book, Johnson recounts tales which are both inspirational and cautionary, and covers characters who were celebrated as having a certain nobility, some with an anti-capitalist 'Robin Hood' morality, as well as others who were reviled as psychopaths. Common to many of the tales is the implication of alternative ways in which individuals and groups could co-exist. Suggestions that Captain Charles Johnson may have been a pseudonym for Daniel Defoe seem plausible, given the resonance of Robinson Crusoe (1719) with the wild and remote island-based, occasionally utopian, backdrops of *The General History*, and with its social critique.

Johnson and/or Defoe paved the way for the 19th-century development of the romantic remoteness and mystery of the pirate, allowing him to be both villainous and noble, as fictionalized by writers such as Byron, Walter Scott, Charles Kingsley, Washington Irving, James Fenimore Cooper, Edgar Allen Poe, R. M. Ballantyne, and Robert Louis Stevenson. In the first half of the 20th century Hollywood made the genre its own, helped by the books of Rafael Sabatini and Howard Pyle. In movies, the pirate became a less ambiguous and rather wholesome swashbuckling figure. Captain Blood, played by Errol Flynn, is one of the defining portrayals of the 20th century. Movie pirates are generally characterized as falsely accused, or pursuing alternative justice; sometimes a dispossessed aristocrat, a rebel with a cause, or essentially any man 'fighting for the right in a world that does not understand the right as he sees it' (Parish 1995). Despite the commercial success and popularity of so many pirate films, the genre eventually lost the wind from its sails. By the second half of the 20th century, film representations, along with TV and comic strip portrayals, veered towards comedy, pastiche and the children's market, e.g. *Captain Pugwash* (1957); *Yellowbeard* (1983); *Hook* (1991). The pirate became emasculated and often a figure of harmless fun. *Cut-throat Island* (1995) is indicative of the creatively exhausted nadir of pirate-themed movies, and is claimed to be the biggest box-office failure of all time,[9] leading to an eight-year hiatus, before Captain Jack Sparrow revived the genre in 2003.

Piracy as resistance to privatization and prejudice

No man will be a sailor who has contrivance enough to get himself into a jail

Samuel Johnson's well-known phrase reinforces the popular notion that for sailors on lawful commercial and navy vessels of the 16th and 17th centuries, many of whom were not voluntary members of the crew, the working life could be cruel, violent, insanitary and miserably remunerated. The prevailing brutality of 'legitimate' labour, combined with the political radicalism of proletarian rebels who failed to gain political voice in the English Revolution, meant that there was no shortage of articulate and passionate individuals who were attracted to the pirate life as an opportunity to be part of an alternative anti-authoritarian and self-determining community. The attraction was supported by a proliferation of pirate-related tales of abundant 'virgin' territories, unspoiled by the tyranny and corruption of church, state and capitalism. Being exiled through prejudice and institutional non-compliance, pirate communities were often relatively tolerant and open-minded on matters of race, religion, class and sexuality. Jo Stanley's *Bold in Her Breeches* (1996) suggests that gender role and behavioural prejudices were also less restrictive in the pirate world.

Being an outlaw of the state did not necessarily imply lawlessness. There are several accounts[10] of pirates having their own well-developed rules, codes and disciplines which were applied with the same rigour as the law. Pirate codes included 'articles'[11] which were a form of social contract considerably more egalitarian than commercial or naval vessels, and which gave crew members a vote, rights to division of stores and plunder and even disability benefits. There was also brutal 'justice' for those who breached the rules. But more significantly (and less depicted in fictional representations) pirate captains were often elected. In such democracy, absolute authority was only asserted during conflict. The details of these accounts of life on board are varied, but 'are unified by the absolutely radical idea that authority depended on consent' (Parker 2009, p. 176).

With democracy, tolerance, respect and discipline not being entirely uncommon features of pirate code, it is not surprising that favourable accounts of alternative societies spring up around pirate activities. Libertalia is one such utopia mentioned in the second volume of Johnson's General History.[12] It was established by Captain Mission in northern Madagascar and is founded on equality of ownership, religious tolerance and a democracy with elected councillors who have a three month term. Madagascar is also the location of libertarian settlement formed by Captain Avery as recounted in the *King of Pirates* (Defoe 1720). There is no way of reliably knowing whether these communities existed in the form recounted, and even if they did, for how long[13] and with what success. For my purposes, their historical accuracy is not so relevant. I am more interested in how positive accounts of pirate solidarity, values, codes and communities remain intertwined with more morally one-dimensional accounts, such as Howard Pyle's description of Tortuga:

> from that spot, as from a center of inflammation, a burning fire of human wickedness and ruthlessness and lust overran the world, and spread terror and death …
>
> (Pyle 2006, pp. 25–26)

The competing narratives are evidence of a long-running rhetorical antagonism between piracy and legality, and between individual freedom and state-authorized power, which has great relevance to current debates over intellectual property reform. *The General History* is much cited for its radical social critique. From a later edition, Captain 'Black Sam' Bellamy, regarded as something of a Robin Hood figure, is cited for his contempt for other captains who 'submit to be governed by the Laws which rich Men have made for their own Security'. He describes those who give in to such authority as being like a 'sneaking Puppy' or as 'hen-hearted Numskuls'. He asserts the justification of his own code by claiming that the only difference between him and the capitalists is that 'they rob the Poor under Cover of Law, forsooth, and we plunder the Rich under the protection of our own Courage' (Defoe 2005, intro., p. 8).[14] This is reminiscent of Cicero's anecdote:

> For it was a witty and truthful rejoinder which was given by a captured pirate to Alexander the Great. The king asked the fellow, 'What is your idea in infesting the sea?' And the pirate answered with uninhibited insolence, 'The same as yours in infesting the earth! But because I do it with a tiny craft, I'm called a pirate: because you have a mighty navy, you're called an emperor.'
>
> (Cited in Johns 2009, pp. 36–37, attributing to Augustine of Hippo)

Similar anti-capitalist sentiment can be found from one of the best-known pirates of the golden age, who has the longest entry in *The General History*: the Welsh captain Bartholomew Roberts, sometimes known as Black Barty. Charles Johnson attributes the following, much-cited, lines to him:

> Damnation to him whoever lived to wear a halter
>
> In an honest service there is thin commons, low wages and hard labour. In this [the pirate life], plenty and satiety, pleasure and ease, liberty and power. And who would not balance creditor on this side, when all the hazard that is run for it, at worst, is only a sour look or two at choking? No, a merry life and a short one, shall be my motto.
>
> (Johnson 1998, pp. 213–214)

The goal of the pirates seems to have been not so much to *accumulate* wealth and property as to *enjoy* it, even if that meant wasting or destroying the possibility of longer-term value, including life itself. According to Cordingley, burying treasure was rare, it being much more common to enjoy their plunder while they could. In this sense, pirates deny the legitimacy of property-owning ideology, and they fight those imperialists and capitalists who would accumulate property and wealth and who would wait for returns on investment through the exploitation and deprivation of others. The words 'balance creditor' seem deliberately chosen, in order to make Roberts' argument by using the double-entry bookkeeping metaphor of the enemy, possibly suggesting Johnson's (or Defoe's) underlying political agenda. As Parker (2009) points out, 'if the sea had been a global commons, it was now subject to enclosure on behalf of states, in turn acting on behalf of an emerging capitalist class' (p. 181). Pirates resisted such enclosure, but having no state or territory, their identity and their capacity to establish a robust and lasting discourse of alternative forms of social organizing was as fluid and rootless as the sea upon which they sailed.

The enduring significance of the pirate

Sympathizing with the golden age pirates may be as romantic and as dangerous as viewing 21st-century Somali pirates simply as brave fishermen protecting their coastline against international privateering and toxic waste-dumping. The habitual violence of pirates, combined with their lack of respect for property law, mean they are destined always to play the role of antagonist in a predominantly capitalist, globalizing narrative. However, their counter-narrative would have lost its connection with the population at large, and would have disappeared a long time ago, if the motives, morality and fairness of the protagonist position were less questionable. The barefaced confidence and notable recent election successes of the Pirate Party[15] in various countries suggest that pirates cannot be regarded as solely historical phenomena. Their resurgence, if indeed they ever really went away, means a strengthening of a counter-narrative which is morally rooted in the protection of the commons from those who would privatize it.

Nowhere is this more apparent than in the construction and protection of the 'virtual commons', which is symbolized, though not limited to, the internet. The irony here is that, by perpetuating a discourse of piracy, which bases its universal jurisdiction on the 'heinousness' of its offences, those fighting piracy inadvertently invite ridicule, and weaken their argument by using a metaphor which has lost its power to shock. Depriving an artist of his livelihood may resonate with the public as an unwholesome pursuit, but it is not the same as robbery with cruelty and violence, especially when the pursuit may sometimes be political rather than commercial, or sometimes motivated more by the desire to create, share, inform and educate, rather than simply to gratify. Moreover, the fear and mistrust of big corporations means that the use of the piracy discourse is more likely to provoke and promote a deeply embedded historical fantasy – that of the pirate as the liberated individual who understandably rejects the constraints of a property-based system which seeks to make money by constraining his freedom, and, in a new 21st-century twist, by exploiting the privacy of his social networking and of his search and browsing history.

13
Enclosing the Commons of the Mind

'This is for everyone' – so tweeted Tim Berners-Lee, inventor of the World Wide Web, at the opening ceremony of the London 2012 Olympics. The message was instantly spelled out in lights around the cheering stadium and transmitted to hundreds of millions of people around the world. The internet as virtual commons, to which every citizen has a right of access, is now a well-established principle which few governments dare to challenge. The complicated and contested part is defining precisely what, amongst all the content and services available on the internet, should be in the public domain, what should be private and subject to protection, and what should be done by governments to ensure fair play amongst all interested parties.

What and where is the public domain?

The first challenge for the digital public domain is in clearly establishing its location and scope in the imagination of the public. Where is this place where the public hold dominion? Indeed which public – national or global? With its goal of exchanging data as efficiently as possible, the internet was not developed in a way which could respect territorial legal and cultural differences. The internet can de-territorialize human activity in that it makes us more receptive to the idea that we are part of a network or a community where geographic and political borders are less clear and less relevant. De-territorialization is a term which can refer to the weakening of ties between culture and place, a phenomenon well illustrated by colonial history where invading nations eliminate the symbols and rituals of a conquered territory and replace them with their own. The 20th-century colonization of the globe with Anglo-American popular culture can be seen as a modern version of de-territorialization which was hugely lucrative for the music and film industries because the process of distribution was within their control. David and Kirkhope (2005) identify the irony, the paradox and the crisis tendencies now arising from that same process for industries which cannot survive without imposing artificial scarcity on an internet which defiantly resists scarcity. This quality of the internet might suggest that the public domain, in practical if not in legal terms, has become a considerably bigger domain in the 21st century. Nevertheless, rights-holding industries are desperate to re-territorialize, and to pursue actions defined within the constructs of the physical world, and to limit the definition of the public domain only to that which falls beyond their asserted rights. They have to change their actions depending on the jurisdiction of the territory they are targeting, as different nation states interpret copyright law differently, with varying degrees of emphasis placed on individual liberty and cultural protection. For these reasons the establishment of a commonly understood definition of the public domain seems a remote possibility.

In his book *The Public Domain: Enclosing the Commons of the Mind*, James Boyle (2008) covers the full range of idiosyncrasies and absurdities of intellectual property law, drawing examples from science, culture and business. He is by no means anti-property or anti-copyright, but he does make a compelling case for more public engagement with the topic to bring equilibrium into the debate about the future of the public domain. He draws attention to the dangers of the field being regarded as an inaccessible, complex and esoteric topic, and of it being left to the lawyers to sort out. The book is a discursive 'call-to-arms' for the establishment of the public domain as a stronger social construct, a Grand Discourse in its own right, not just a residual 'catch-all' term for everything that is not enclosed by property rights.

The analogy implicit in his enclosure metaphor, between the restriction of access to common land and of access to a network for the sharing and further cultivation of intellectual, scientific, creative, social and commercial activity, is more intuitive and compelling to some people than it is to others. One of the limitations of the effectiveness of the analogy is the relatively narrow but essential activities one might undertake on the common land (grazing, cultivation) compared with the enormity and diversity of activities which occur via the virtual commons. Another limitation is the difference in clarity and consistency of the 'fencing'. When common land in England was fenced off and passed into private ownership (over the course of hundreds of years), its boundaries were normally visible, understandable (though not always socially acceptable) and relatively easily enforceable. By contrast, there is inconsistency and confusion about the rules, boundaries, rights and privacy within the virtual commons.

Where the internet-as-commons metaphor does work effectively, however, is in providing a mutually recognizable historical reference point in the ongoing economic argument for and against 'enclosure', which is effectively just a word to describe the creation of private property. Enclosure of common land progressed as widely as it did because it demonstrated the productive possibilities of land when managed by a private owner who was incentivized to manage it efficiently, knowing that any investment he makes in the land would be protected. The argument relies on a view of human social relations which assumes that without some incentivizing and regulating mechanism, individuals are either too self-interested, or too lazy, to optimize the productive use of the land, i.e. the land will be either ruinously over-grazed, or wasted. The extent to which the increased productivity from enclosure made for a better or fairer society remains a highly contentious topic, but the story of the enclosure movement being an efficient solution to the risk of the 'tragic' consequences of the commons has gained strong currency.

It is worth digressing for a moment on the 'the tragedy of the commons' because it is an interesting and loaded phrase. It is interesting, and indeed frightening, to consider why the proposition that the commons is inherently 'tragic' became so readily accepted. The source of the phrase is commonly attributed to the title of an article in the journal *Science* by bio-scientist Garrett Hardin (1968). Echoing Malthusian concerns about population growth, Hardin tells a tale of herdsman over-grazing their pasture, and claims that population growth falls into the category of problems with 'no technical solution' (p. 1243), due to the *remorseless inevitability* of the self-interest of the human condition. I highlight the words *remorseless* and *inevitability*, which Hardin repeats in his article, to emphasize how his language draws on a discourse of determinism which is tragic in its denial of hope of human salvation without bold political intervention. He calls for a 'recognition of necessity' (p. 1248) that the notion of 'commons' must be abandoned as the population increases, in particular abandoning the freedom of humans to breed. Hardin's paper was published the same year that the Indian government set a goal to reduce the birth rate by 45% within a decade, largely through a programme of sterilization, and ten years before China introduced its one-child policy. These drastic and desperate social policies are indicative of a prevailing notion that (a) the planet's resources are finite,[1] and that (b) without intervention, individuals are incapable of behaviour which is in the long-term benefit of the planet. Forty-five years later, the inevitability[2] of population growth is still one of the largest global concerns, but there is broader acceptance that solutions to population growth should be limited to more dignified and constructive policies, such as the education of women and their wider integration into the workforce, the reduction of poverty and access to family planning resources. The focus of contention and sensitivity has shifted from Hardin's restriction on the 'commons of breeding', to the questioning of the 'commons' of a universal consumption entitlement which is modelled on the voracious 'developed' world.

The tale of the tragedy of the commons may seem to be a digression, but I mention it because it does indicate the power of master-narratives to feed ideological conflict between elite-capitalist and left-wing or environmental views. The tragedy of the commons has alternatively been described as a 'pernicious myth' (Lewis 2012) on the grounds that it ignores the 'triumph of the commons' as a sustainable communal option for land usage. It promotes a fear-inducing image of scarce resources and a global race to mutual destruction; a scrambling free-for-all which puts population growth at the top of the list of threats to planetary despoliation, and conveniently *demotes* the cultural and economic risks of the international development agenda of Western-style education and privatization. The phrase 'free-for-all' captures perfectly the paradox which can be used by either side of the argument: on the one hand it implies open access and a level playing field, and on the other, anarchy and aggressive self-interest. Developed nations have a huge middle class whose security and values are perceived to depend on protection of 'capital' – not just homes and savings, but also intellectual/educational, cultural and social capital which preserve the prevailing social order. Such capital is rooted in a predominantly enclosed system of property rights, in one form or another, which several generations have grown up with as a robust principle of social organizing. Relaxing the boundaries of property rights and extending the commons may therefore be associated with, at best, dilution in value, quality and rigour, and at worst, anarchy and lawlessness.

The tendency for many people to lean towards these negative and frightening aspects of 'free-for-all' has been described as 'openness aversion', or 'cultural agoraphobia' (Boyle 2008, p. 231). To counter such a tendency, Boyle argues that 'we have to "invent" the public domain before we can save it' (preface xv). There are however obstacles to doing so. The first is the inaccessibility of intellectual property law, due to its overwhelming combination of technical complexity and enormous range: patents, copyrights, trademarks, all operating across science, technology, healthcare, education, business and culture. There are many issues with which the public and general media do engage. These include: patents on human genes; pricing of educational textbooks and scientific journals; lawsuits against whistleblowers who have infringed copyright in the interests of uncovering more serious wrong-doings; affordable access to HIV drugs in poor countries; spurious copyright claims or excessive constraints by the entertainment and publishing industry over 'fair use'; impossibly onerous clearance processes which stand in the way of cultural production; and excessive penalties against casual downloaders. The list goes on, but these are often seen as fragmented and isolated issues, rather than being indicative of a larger and connected social problem of intellectual property which could be coherently understood and generically articulated.

The diversity of these issues presents an obstacle to the establishment of commonality, solidarity and political voice amongst the very different communities which care about them. Can a bio-scientist, a librarian, an aspiring musician, a developing-world government minister, a documentary film-maker, a progressive teacher, etc.... practically come together to help promote a new discourse of the public domain? According to Boyle, there are lessons to be learned from the growth of the discourse of 'environment' which emerged over the second half of the 20th century, despite facing similar problems of scattered communities which had a huge diversity of interests and concerns, all demanding more political equilibrium. One can imagine that a campaigner for the reintroduction of the beaver to Scotland might not have much to say to a Chinese prosecutor involved in a case of the corporate bribery of an environmental pollution inspector. Neither of them would feel an *esprit de corps* with the Head of Renewables at an investment bank. Similarly, within what is now generically recognizable as the *environmental movement*, there can be some ferociously polarized positions, such as those over whether animals and humans have equal rights, or whether the term 'international development' is intrinsically constructive or destructive. But what continues to loosely tie all these disparate agendas together is a shift towards the perception of a common self-interest, and an accompanying coherent discursive repertoire to make the whole environment discourse seem greater than the sum of its parts. The environmentalists' repertoire drew from the science of ecology, which transcended the parameters of its scientific field to become a metaphor which reinforces the importance of equilibrium and the sustainability of diverse interactions and dependencies. It also drew from economics to reveal that traditional accounting standards fail to capture all the costs of market-driven activity, especially the costs to a broader set of stakeholders, including Mother Nature. This then spawned a whole new language of corporate social responsibility, of which environmental impact is one element regularly measured and reported by corporations, often over-simplifying the environment as something which is a direct beneficiary of a reduction in a company's carbon footprint.

The 'environment', in its now common sense usage as 'the natural world... esp. as affected by human activity' (OED, definition 'd'), grew in influence from the 1950s on, largely due to the compelling narratives and rhetoric as rehearsed and practised by the communities and networks which sought to raise awareness. Such has been its success that it is now difficult to contemplate the word *environment* without simultaneously conjuring up images of the Earth and its vulnerability to destruction by mankind. This very powerful social construct of the environment has literally been talked into being, albeit with the help of pictures, films and graphs, to move the complex and inaccessible fields of environmental sciences, including their economic and political dimensions, into an engaging and popular set of stories that we tell ourselves to help us make sense of the planet's ecological complexity.

The narrative challenge of the public domain

Despite the diversity and complexity of the issues, the overall *raison d'être* of the environmental movement is intuitive to most people. Mother Nature's story is quite simple and powerful: if individuals, state-institutions and commercial organizations are not measurably and restoratively accountable for their pollution, depletion, or unsustainable, non-renewable resource consumption, then we can expect permanent damage to the planet, and we will all suffer as a consequence, sooner or later. Antagonists in the narrative vary in degree: rapacious and unchecked capitalist corporations; the Chinese government with its hubristic economic miracle and aspiration to be the world's superpower; or simply mankind's abandonment of community, drawn towards the rocks of destruction by the siren of technological 'progress' and the promise of endless gratification through consumption and novelty. Protagonists include Mother Nature, political activists, scientists (at least those who are free from corporate or excessive national interest), and responsible citizens. Competing plot variations do cloud the issues to some extent – are resources finite? When, if ever, will peak population be reached? Will technology that we have not yet invented save us? Is climate change real, and if it is, what can we practically do about it? – but at the highest level, the master-narrative of the environment has become a compelling counter-narrative to the discourse of globalization. Its only serious competition is the International Developer's seductive tale of social justice and the elimination of poverty through education and economic growth, to which Mother Nature's tale is sometimes a thorn in the side, but that is a subject for another book.

Protection of the intellectual and cultural public domain does not, at least not yet, lend itself to outrage and fear-mongering, and anxiety about the world our grandchildren will inherit, in quite the same way as images of polluted rivers, smog-choked cities, devastated rainforests, scorched earth and stranded polar bears. Trying to get broad public engagement with the inequities of intellectual property law in the digital age is a tough challenge. Actually, there is plenty of engagement on individual topics; the problem of engagement only really exists when the topic is discussed in the aggregate or in the abstract. The terms *public domain*, and *intellectual property* do not really connect with people at an intuitive or empirical level. Tell people that the vast majority of 20th-century culture is commercially unavailable,[3] and many, if not most, will give a shrug of the shoulders and suggest that maybe it isn't any good. People need examples, metaphors, stories, and a coherent framework which helps them intuitively link the examples. Then they might see that there may be a social imbalance in intellectual property law, the consequences of which (intended or unintended) they might actually care about quite deeply if they understood it better. For now at least, the master-narrative of the commons has *not* been unequivocally established.

There is no shortage of individuals and organizations who have been championing the public interest in response to new technologies and to try to redress a disequilibrium which, it is argued, remains in favour of content-owners and corporate interests. These are variously and informally aligned under umbrella 'movements' such as Access to Knowledge (A2K) and Open Access (OA) amongst many others, each with their own priorities. So, for example, the A2K network is run by Consumers International and covers a broad range of consumer interests, including privacy and surveillance. Open access has its roots in software development but OA is now mostly recognized to refer to the context of scientific and academic publishing, with aims to make scholarly journal articles freely accessible for consumers by changing the traditional concepts of the publishing business model. Public Knowledge and the Open Rights Group are active public interest groups in the US and the UK respectively who campaign on a wide range of issues relating to freedom of expression, privacy, innovation, creativity, consumer rights and preservation of an open internet.

To go into detail on all such champions of public interest would be a large and distracting undertaking, but it is worth mentioning a couple of pioneers who became 'centres of gravity' in the emerging alternative discourse of intellectual property. The Electronic Frontier Foundation, with its motto 'defending your rights in the digital world', is one such pioneer. It was founded in 1990 and its roots are in concerns for the protection of freedom of speech, rather than in the reform of intellectual property rights. It is almost 20 years since its co-founder John Perry Barlow wrote a seminal piece for *Wired* magazine called *The Economy of Ideas* (1994) which massively boosted an alternative discourse of intellectual property. He called for a new social contract for 'cyberspace' (a term he coined) where protections should be based more on ethics and technology than on practically unenforceable laws and rules. He also proposed a new 'taxonomy of information' to differentiate the essential characteristics of 'unbounded creation' in the digital age from previous forms of industrial property. This new taxonomy regards information as an activity, a verb rather than a noun, a life form (which wants to be free), and as a relationship. As a source of discourse genealogy, *The Economy of Ideas* remains relevant and valuable today, and will resonate with people even more after two decades of being challenged, derided, expanded and built upon by thousands of articles and blogs. *Wired* magazine itself has been a very significant source of new discourse, and the ecology background of its pioneering editors (Kevin Kelly and Stewart Brand) is indicative that leveraging tropes from the environmental movement can be effective.

The second discursive 'centre of gravity' I want to mention is the US law professor Lawrence Lessig. His 1999 book *Code: and Other Laws of Cyberspace*, updated in 2006 as *Code Version 2.0*, was a cautionary argument against the then common view of techno-utopians like John Perry Barlow, that cyberspace was beyond the reach of government regulation. That view is best exemplified by Barlow's 1996 Davos speech, or rather declaration, which rejected any government claim of sovereignty over cyberspace. Lessig is equally concerned and passionate, but more politically pragmatic in his scepticism of government intervention on the internet. He is a strong opponent of the influence exerted by large media companies over policy-making, maintaining that entrenched commercial interests mean that the internet is ever more tightly regulated, even hegemonic, thereby constraining its innovative potential. In this respect, he argues that recent extensions to copyright law are fundamentally unconstitutional in favouring rights-holders over the public interest. Practically speaking, his focus is the unnecessary restrictions on the usage of copyright material by consumers and creators in their own endeavours. The irony implicit in Lessig's argument is that the old intellectual property system, which was originally designed to promote culture and creativity, has become, through corporate protectionism, unnecessarily restrictive and now stands as an obstacle to the creative possibilities of new media and technologies. His analogy that the law has been re-written so that 'no-one can do to the Disney Corporation what Walt Disney did to the Brothers Grimm' (Lessig 2002) is widely quoted. In his explicitly titled *Free culture: how big media uses technology and the law to lock down culture and control creativity* (Lessig 2004), he expounds the alternative copyright system of Creative Commons (www.creativecommons.org), of which he was a co-founder in 2001. Creative Commons offers creators an alternative to the traditional 'all rights reserved' by offering content producers a free and flexible licensing mechanism of only 'some rights reserved', thereby unlocking the creative output of hundreds of millions of works[4]. As an alternative licensing system which tries to at least partially reconcile the old and new world copyright conflict, and which has been recognized and adopted by so many people worldwide, Creative Commons really is a remarkable achievement, the historical significance of which may not yet have been fully recognized.

As influential as Barlow, Lessig and many other articulate visionaries and scholars of the digital age have been, a clear master-narrative of the public domain remains elusive. One of the challenges is that there is little consensus on the denouement of the plot or on the characters, especially the protagonist. The techno/cyber-utopian camp had a very promising revolutionary plot opening of 'a new Home of Mind, naturally independent of the tyrannies' which the 'weary giants of flesh and steel' seek to impose.[5] Barlow's antagonist was clearly, if broadly, outlined, but the protagonist was a hazy and undefined 'we', implicitly huge and homogenous in its desire that knowledge should break free from the control of an industry–government alliance. His declaration is rhetorically powerful but lacks a narrative framework on which others, including policy advocates, might sensibly build. Another weakness is that, because the majority of peer-to-peer sharing does not involve the creation of new work, the discourse has been vulnerable to being dismissed as a discourse of 'dotcommunists' or 'freetards', the latter being a derogatory term for 'a person whose ability to make rational decisions is impaired by the possibility of receiving something for free'.[6]

At first glance, Lessig's antagonists of big media and government look the same as those personified and demonized by the cyber-utopians. However, Lessig, like Boyle, is more pragmatic, recognizing that an anarchic dismantling of copyright is not a desirable end, as it fails to reflect the essential societal dilemma of promoting and protecting cultural exchange and production. His plot is therefore more about striving to moderate, and to achieve a better balance of stakeholder interests. The aim is to reduce the way in which 'all rights reserved' copyright thoughtlessly locks up all creative works as if they were finished products with no metamorphic or reincarnated afterlife, just in case they might be valuable later, or in case other people's derivative adaptation of them might dilute their exclusively controlled asset value. Lessig's aim might be achieved by, amongst other measures, significantly shortening copyright term, though making it renewable, and losing rights where owners do not make the works available. These are strong arguments but, being moderate and balanced solutions for problems which are diverse, complex and contested, they are often expressed in subtle, intellectual and esoteric language. They do not lend themselves to having polarized or colourful protagonists and antagonists, nor to having an accessible master-narrative which can appeal to a wider public through anger or through fear-inducing tales of life-diminishment and injustice.

Having apparently recognized the futility of striving further for copyright reform under the current system of lobbying politics, Lessig himself has changed strategy in recent years, and has taken on a more powerful and emotive antagonist: the US political system itself and how it has become corrupted such that it would be no longer recognizable to the Founding Fathers, nor to the Framers of the Constitution. The scope of this book prevents further analysis of this development, but Lessig's 2013 TED Talk[7] is worthy of viewing as a narrative masterclass, illustrating the self-protecting nature of the funding and voting system and how it has become hard-wired to maintain the status quo and resist reform of any kind from left or right.

The public domain has some way to go before it achieves the same kind of universal recognition and common sense acceptance of its need for protection. In the meantime the narrative landscape remains one where private property prevails. Unless there is a radical shift in public perception towards the alleged injustice, inefficiency or corruption of intellectual property law, then originators and their corporate patrons and protective-curators will continue to tell more successful tales than the liberal-curators and disseminators, the more radical of whom will remain characterized as troublemakers, thieves or pirates.

In the next chapter, I look briefly at the narrative history of copyright law in the context of the contested legitimacy of authorship and its protection, and offer some thoughts about the current status of calls for the reform of copyright.

14
The 300-Year War of Copyright

> If one needs an army of lawyers to understand the basic precepts of the law, then it is time for a new law
>
> (The Next Great Copyright Act,[1] speech by
> Maria Pallante, United States Register of
> Copyrights, 4th March 2013)

> My real message was, let's not just put enforcement bills on the table and then be surprised when the public doesn't understand. We've got to tell a better story.
>
> Maria Pallante (December 2013)

Two years into her role as Register of Copyrights and Director of the US Copyright Office, Pallante saw fit to make a speech calling for a comprehensive revision of copyright law. She recognizes that progress is likely to be slow due to the complex and arcane provisions of statute, the intensity with which interested parties make their views known, and the public's confusion, if not aversion, on matters of copyright (p. 5). Nevertheless, her call is an acknowledgement that the technological and social changes of the 21st century mean that the law is not doing its job and risks losing its moral authority. Merely tinkering with amendments to existing statute is highly unsatisfactory. A better story needs to be told.

The contested history of copyright law

Some may regard Pallante's call as an invitation to open a can of worms, but it is a can which has perhaps never been satisfactorily closed. Copyright has been a contentious topic for more than 300 years, and the debate erupts ferociously from time to time, especially when technology changes, or when the balance of power between trade, state, and culture becomes unstable. This is because the topic has profound implications for reassessing fundamental conceptions of the creative process, of human rights and of social organization. In one short chapter, I cannot possibly do justice to the vast amount of scholarship and polemic on the topic, so I will excuse myself in advance against complaints of oversimplification, or of arousing curiosity in the reader which may only be satisfied from delving into other sources.

There is a certain degree of consensus in basic principle that some form of economic protection of creative labour can act as an incentive to promote innovation which is good for society. Thereafter, philosophies diverge fairly quickly about how the conflicting interests inherent in the creation of artificial scarcity should be balanced. I am going to return to the assumption that narrative is a way in which intractably polarized dilemmas can be more manageably conceived and negotiated. At certain times, all parties to an issue become fully engaged in a debate or dispute and can easily recognize the simplifying narrative mechanisms. Once one story emerges as more compelling, it gets repeated so often that it becomes culturally embedded as a masterplot, sometimes even in statute, and it lies relatively dormant and uncontested until something comes along to resurrect the contest. Digital technologies and the internet have been such a catalyst: resurrecting the storytelling contest because much, if not most of the technological and social infrastructure for expressing, sharing and developing the products of human thought, creativity and ingenuity has changed beyond what could possibly have been imagined when the foundations of today's statutes were laid.

My simplification of the polarities of the copyright dilemma is framed in Figure 14.1. Some of these elements are less polarized than others, but I find it helpful to group them in this way as most, if not all arguments made in the 21st-century digital storytelling contest can be mapped to these discursive repertoires, directly or indirectly.

Most of the terms in Figure 14.1 will by now be familiar to the reader, but before I comment on them further, I want to pick up where we left the pre-history of intellectual property in Chapter 12, with the demise in 1695 of the monopolistic licence granted to printers and booksellers. For the following 15 years, there was widespread printing without the consent of the author, and beyond the control of the printers and booksellers. Desperate for some reinstatement of structure which they could control, and recognizing that a discourse of authorship was on the rise, the booksellers de-emphasized talk of their own rights. They strategically positioned themselves as logical stewards of authors' rights. Being able to campaign for their own interests under the banner of authors'

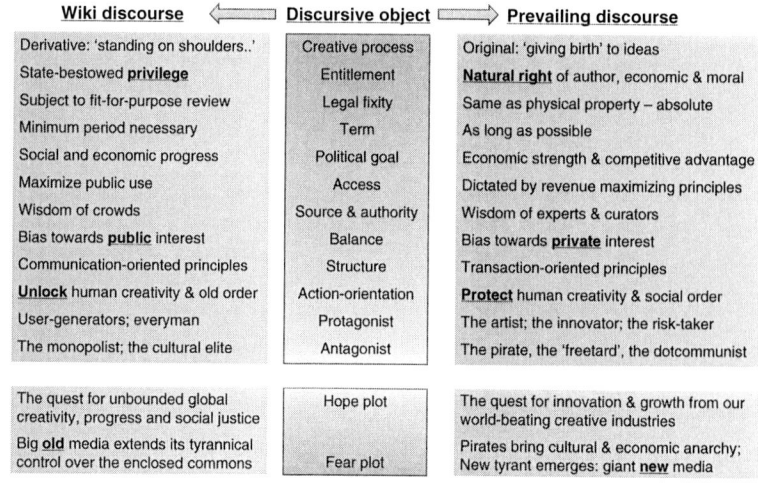

Wiki discourse	Discursive object	Prevailing discourse
Derivative: 'standing on shoulders..'	Creative process	Original: 'giving birth' to ideas
State-bestowed **privilege**	Entitlement	**Natural right** of author, economic & moral
Subject to fit-for-purpose review	Legal fixity	Same as physical property – absolute
Minimum period necessary	Term	As long as possible
Social and economic progress	Political goal	Economic strength & competitive advantage
Maximize public use	Access	Dictated by revenue maximizing principles
Wisdom of crowds	Source & authority	Wisdom of experts & curators
Bias towards **public** interest	Balance	Bias towards **private** interest
Communication-oriented principles	Structure	Transaction-oriented principles
Unlock human creativity & old order	Action-orientation	**Protect** human creativity & social order
User-generators; everyman	Protagonist	The artist; the innovator; the risk-taker
The monopolist; the cultural elite	Antagonist	The pirate, the 'freetard', the dotcommunist
The quest for unbounded global creativity, progress and social justice	Hope plot	The quest for innovation & growth from our world-beating creative industries
Big **old** media extends its tyrannical control over the enclosed commons	Fear plot	Pirates bring cultural & economic anarchy; New tyrant emerges: giant **new** media

Figure 14.1 The contested narrative of copyright

rights was effective, and indeed remains an effective lobbying strategy for content-owners today.

The Act commonly referred to as the first copyright legislation was the Statute of Anne (1710), but in fact the word copyright does not appear in it, and only emerges 25 years later. The statute was titled *An Act for the Encouragement of Learning, by Vesting the Copies of Printed Books in the Authors or Purchasers of such Copies, during the Times therein Mentioned.* In its opening line it refers to 'the very great detriment, and too often the ruin of them [authors or proprietors] and their families', caused by non-consensual printing. Though not obvious from its title, the Act represented more of a victory for booksellers and publishers than it was for the recognition of authors' rights, because in practice it legitimated the publishers' monopoly and destroyed any right (if one ever existed) to authors' perpetual copyright in their published work. The term awarded was 14 years, which was renewable for a further 14 years if the author was still alive. Works already in print when the Act was passed were given a protected term of 21 years.

Copyright term 1710–2013

As a pragmatic resolution of an unstable situation, the various interested parties were reasonably satisfied with the Statute of Anne and the arguments were relatively settled for the next 20 years, until the first copyrights began to expire. Then began the 'battle of the booksellers', a sustained effort to extend copyright term and argue that a common law right in perpetuity pre-existed the Statute of Anne. A number of legal cases are often cited as the discursive focal points, such as Tonson v. Collins (1762) and Millar v. Taylor (1769), culminating in the ruling in the case of Donaldson v. Beckett (1774) which held that the Statute of Anne restrained or removed any perpetual author right which might have existed in common law. The ruling upheld the 14-year copyright term, but did not definitively close the debate on the possibility of longer terms. Over the following century, economic and philosophical arguments were heatedly exchanged, not just in Britain, but also in France, Germany and the US. In Europe, Romanticism fuelled the perception of author as original genius, and the metaphor of his output as property became more culturally and commercially embedded without much alternative championing of the public domain from the 'hollowed out'[2] public sphere. Publishers had little objection to the strengthening of author property rights, as rights were easily acquired or relatively cheaply licensed to corporations, giving them exclusive control. Consequently, by 1842 the copyright term had lengthened to a 42-year minimum, or the life of the author plus seven years. By the time of the Berne Convention (1886), there was some acceptance of the benefits of laws which could be internationally harmonized and binding. The Berne Convention stipulated that parties to it must have a *minimum* term of author's life plus 50 years. In 1996 the UK term was extended to author's life plus 70 years to harmonize with the European Union. In 1998 the US Copyright Term Extension Act[3] also extended to life plus 70 years. So in practice this means that today, for a 20-year-old taking a photograph, writing a song, a blog, or a computer program, if she lives until she is 90, then she and her dependents will by default enjoy 140 years of copyright protection.

Historiography of copyright law

In the introduction to *Privilege and Property, Essays on the History of Copyright*, Deazley et al. (2010) invite more work on the historiography of copyright, i.e. a study of the narratives about the changing construction of copyright history at various periods, asking such questions as how objects of inquiry and primary sources are identified, the causes and effects in the evolution of the norms of copying, the extent to which those norms are served, dictated or ignored by the law, and 'which justificatory goals are served by historical investigation' (p. 3). This might seem an esoteric and overly intellectual exercise, but it is relevant if one wants to unmask the dominant narratives which might potentially stand in the way of the kind of reforms urged by Maria Pallante.

One of Pallante's predecessors at the US Copyright Office said in 1974 that it was 'of debatable significance' that copyright originated in England centuries ago:

> This was a period of great religious ferment and political unrest during which witchcraft and devil-worship were at their height, and repressive measures against all forms of heresy were widespread. [...] I don't agree with the charge that copyright originated as a marriage between tyranny and greed, arranged by the devil. [...]...the first copyright statutes were based on a rejection of autocratic repression and monopoly control and upon a new recognition of individual liberty and the human rights of authors.
>
> (The Demonology of Copyright, Ringer 1974, p. 12)

If the Statute of Anne had taken force in 1695, immediately after the lapse of the Licensing of the Press Act, these comments might ring truer. But the statute came into force after the experience of 15 years of open and unregulated publishing activity, in order to restore quality control and integrity to the generation of knowledge by authenticating attribution, and to provide a living for authors and publishers. Ringer invokes a frightening and slightly anachronistic 15th- and 16th-century image of witchcraft and devil worshipping to rhetorically support an interpretation of copyright history which is more aligned with a European 20th-century construct of the creative process and of individual rights and liberties. I quote this as an example of why one should approach historical copyright references (including any I make in this book) with a critical eye.

How much *authority* do *authors* have?

Returning to Figure 14.1, the first and perhaps the most profound oppositional construct to be reconciled by copyright is that of the creative process itself. What aspects of the process of creation entitle an author or an artist to own, or exclusively to control, the product of his or her labours? This must be addressed as a social and philosophical question before it can be considered as a legal one. There are two popular metaphors which are useful to describe the alternative views: 'standing on the shoulders of giants', and 'giving birth to ideas'. Both metaphors can be traced back many centuries, but as it is not contentious to say that both are culturally embedded, it is not necessary to offer a rigorous genealogy here.

The first is a metaphor which presumes that art and invention are built on what went before. It is often an expression of humility. Its first appearance may be in 1159 as an analogy: 'we are like dwarfs on the shoulders of giants, so that we can see more than they'.[4] Other notable versions include Isaac Newton, Samuel Taylor Coleridge and Stephen Hawking. It is the motto for Google Scholar and regularly referenced by the free software movement. It is also engraved on the British two pound coin. In more recent popular culture, it can be found in the movies *Jurassic Park* and *The Social Network*, and in the title of an album by British rock band Oasis. Philosophically, this view of art, creativity and innovation is that it is largely derivative, emerging as a craft, a product of dedication to learning from others and from practice, and only occasionally punctuated by flashes of untraceable originality, insight or 'genius'. It may not be in question that great artists possess, at birth or through early nurture, extraordinary talents which make their commitment to their craft more fruitful than others, but Mozart is unimaginable without the influence of Bach and Haydn, or Lady Gaga without Madonna and a host of others.

Nietzsche took a slightly different view on the dwarf-giant analogy at least as regards philosophy, seeing the ancient Greek masters as monolithic, devoid of conventionality, and in magnificent solitude: 'each giant calling to his brother through the desolate intervals of time. And undisturbed by the wanton noises of the dwarves [academic scholars] that creep past beneath them' (Pearson and Large 2005, p. 103). Yet even this metaphorical usage does not undermine the view that the vast majority of what passes for original expression is derivative.

In contrast to the idea of the creative process being mostly derived from what has gone before, the alternative view is that of the birth of an original idea. The assertion of an author's unique origination and ongoing responsibility for the integrity of his words is well expressed using the birth metaphor, e.g. the brain-child, though if the author is making a maternal claim, rather than a parthenogenetic one, one might argue that the seed of an idea came from elsewhere, i.e. an earlier dissemination. I cannot trace the first usage of birth as a metaphor for the process of creation or invention, but it undoubtedly pre-dates the English language. In the context of this book, Milton is a useful example. As well illustrated by Rose (2010), the dominant metaphor of *Areopagitica* is the representation of books as living persons, the progeny of their authors:

For books are not absolutely dead things, but do contain a potency of life in them to be as active as that soul was whose progeny they are; nay, they do preserve as in a vial the purest efficacy and extraction of that living intellect that bred them. [...] ...good book is the precious life-blood of a master spirit...

(Milton 2006, pp. 14–15)

Areopagitica (1644) is a seminal defence of the freedom of speech against the oppression of censorship rather than a claim for protection of intellectual property, a concept which would not emerge until over a century later. Nevertheless, positioning the author as progenitor, rather than as dwarf, does provide a stronger case, socially and philosophically, for establishing stronger and longer authorial rights. Milton himself does not make the direct link from progenitor rights to property ownership rights. The link is more clearly made by Diderot in 1763 in his *Letter on the book trade* (see Rideau 2008).

Of course ownership has more than one dimension of meaning. It implies possession which one which can enjoy, but it also comes with the heavy responsibility of accountability. To own a text means to own up to having written it. In late 18th-century revolutionary France, the character and integrity of the author, and the transparent authenticity of authorship, were deemed to be of the utmost importance. Preserving textual attribution and integrity carried as much, if not more, risk than reward, and hence the French are often still perceived to have a continued higher regard than most nations for an author's moral rights. Peifer (2010) points out that the German copyright history is also centred on the personality interests of the author and on his role of information broker for the benefit of society. He believes that the 18th-century 'battle of the booksellers may be told as a story of the balance between a property function and an authenticity function' (p. 350).

The wording of the first US Copyright Act (1790) was essentially the same as the Statute of Anne but the antipathy towards any legal construction of natural or property rights of an originator is famously expressed by one of the Founding Fathers, Thomas Jefferson. Its rhetorical power has at least as much resonance for cyber-utopians today as when it was expressed during the industrial revolution. It is therefore worth citing an extract at length:

If nature has made any one thing less susceptible than all others of exclusive property, it is the action of the thinking power called an Idea, which an individual may exclusively possess as long as he keeps it to himself; but the moment it is divulged, it forces itself into the possession of everyone, and the receiver cannot dispossess himself of it. [...] He who receives an idea from me, receives instruction himself without lessening mine; as he who lights his taper at mine, receives light without darkening me. That ideas should freely spread from one to another over the globe, for the moral and mutual instruction of man, and improvement of his condition, seems to have been peculiarly and benevolently designed by nature, when she made them, like fire, expansible over all space, without lessening their density in any point, and like the air in which we breathe, move and have our physical being, incapable of confinement or exclusive appropriation.

(Jefferson 1813)

Jefferson's letter is actually on the question of a patent[5] rather than a copyright, but by the late 18th century, literary authorship and scientific invention were becoming talked about as similar processes, therefore deserving similar rights and protections. In principle, copyright protects the *expressions* of literary and artistic works, rather than ideas; patents aim to protect new and useful *ideas*. The distinction becomes even more blurred and less intuitive in the 20th century: for example, computer programs are given the status of literary works.

Patenting in one form or another had been around for centuries but had become outdated, inefficient, and arguably stood in the way of industrial progress. It was a licence to litigate, and only offered protection to those who could afford to pay huge costs for complex and often incompetently judged legal cases which could run for years with the outcome being something of a lottery. According to Johns (2009) the Victorian campaign against patenting, which expanded to embrace copyrights also, remains to this day the strongest ever undertaken against intellectual property. Whereas authorship was the domain of the educated elite, the patent debate introduced an element of class prejudice which was rooted in the perception that almost anyone could invent. The inventor was rarely a hero but more often a skilled workman who might simply have the good fortune to be in the right place at the right time, when seeds of invention blowing around in the air were ready to germinate:

in the intellectual world, fitness of time and circumstances promptly calls forth appropriate devices. The seeds of invention exist, as it were, in the air, ready to germinate whenever suitable conditions arise; and no legislative interference is needed to ensure their growth in proper season. (Attributed to inventor and industrialist Sir William Armstrong)

(Johns 2009, p. 269)

Those who argued for abolishing the patent system (a community which included the revered engineer Isambard Kingdom Brunel) argued that science and engineering evolved gradually, methodically and communally. Simultaneous or rival inventions were common, and the patent system was therefore vulnerable to opportunists, schemers and gamblers who could themselves enclose the scientific commons and constrain industrial progress. The battle can be seen as quite polarized between industrial heavyweights (who had the capital and organizational infrastructure which gave them effective monopolies without the need for patents) and a growing class of expert and innovative engineers. At one point, the patent battle was very finely balanced and could have gone in favour of the abolitionists, but a change of government from Liberal to Conservative in 1874 and growing momentum towards including patents along with copyrights in international law (as enshrined in the Paris Convention 1883 and the Berne Convention 1886) led to a reform of the UK patent law rather than its abolition.

Although one should be careful not to generalize too much about the parallel evolution of patent law and copyright law, the eloquent expressions of Jefferson and Armstrong are good examples of the view expressed by Deazley et al. (2010, p. 4) that particular intellectual property laws have come to be associated with distinct philosophical tradition: the US and the UK being public-interest-oriented, or utilitarian, versus France and Germany which are regarded as more originator-centric, and concerned with the preservation of the integrity of an originator's personality and moral intent. The French were heavily influential in the Berne Convention which embodied their concept of author rights, *droits d'auteur*, which distinguish between an economic property right and a moral right. The moral right includes the right to claim authorship of the work and to object to any distortion, modification of, or other derogatory action in relation to the work which would be prejudicial to the author's honour or reputation. The fact that the US did not sign up to the Berne Convention until 1989 may be seen as indicative of a certain discomfort with the potential constraints to open exchange implied by such an enshrinement of moral rights, and, more cynically, the potential impact it has on the balance of power between original author as rights-owner and corporation as licensed rights-holder.

The validity of the author-centric tradition was challenged in the second half of the 20th century. It began in France with post-structuralists Roland Barthes (*The Death of the Author* 1968) and Michel Foucault (*What Is an Author?* 1969) who believed that the presumption of authorial *authority* imposed problematic limits on texts. Texts, in the post-structuralist view, are better understood as existing independently of their authors. Barthes preferred the term 'scriptor' to dissociate author from authority. The scriptor is then someone who provisionally assembles words which already come loaded with a plurality of meanings from prior usage. Meaning, and assertions of 'truth' claims, are unstable, and subject to multiple interpretations within each reader context. In this view, the reader creates the meaning as much as the writer. Martha Woodmansee's *The Genius and the Copyright* (1984) and Mark Rose's *The Author as Proprietor* (1988) are also well-cited works amongst many critical studies which have recognized that, within a long view of the history of knowledge, the presumed significance and *authority* of the author, and his proprietorial relationship with the text, is a relatively recent construct which may be less robust than is commonly accepted.

The numbers of people who aspire to learn, to collaborate, to share, and to express themselves through cultural texts of one form or another, has increased rapidly in the last 50 years. This was first through massively expanded access to education, and then stimulated beyond prior imagining by the cheap and easy technologies of home and mobile computing and global communications. Understandably the law has been challenged to adapt to these phenomena. Yet despite the late 20th-century critical questioning of the significance of individual authorial identity, the old revered qualities of singular authorship and authority, whether in literary, musical or other forms of expression, still have a secure anchoring in the legal protection of cultural production which is in favour of an original creator (and by representation, the rights-holding corporation).

In an attempt to gain some perspective on whether the unchanging institution of copyright is really a problem or not in the digital age, it is useful to take a much longer historical view of human knowledge-creating processes. In doing so, I invite the reader to reflect on whether post-Gutenberg practices represent social progress which must be preserved, or whether they have run their course, being something which in time will be regarded as a historical anomaly which empowered a certain sector of mankind.

Literary authority versus new forms of orality

The aforementioned long view of Western knowledge and authority goes back at least as far as the 4th-century BC, when Socrates (in Plato's *Phaedrus*) tells the tale of the Egyptian God Theuth, and his debate with King Thamus on the propriety and impropriety of granting the Egyptian people the benefits of writing. On the one hand, the technology of writing extends memory and wisdom. On the other hand it will 'create forgetfulness in the learners' souls, because they will not use their memories...they will appear to be omniscient and will generally know nothing.'[6] Hence began centuries of a fragile alliance between wisdom and rhetorical eloquence, and a debate about whether true knowledge can live in static, unresponsive text, or whether it only emerges dialogically from a relationship between active human minds through spoken discourse.

We are born with the neural wiring for speech, but reading and writing are less 'natural' skills only acquired through education and through technologies such as the alphabet. Around one billion people today remain illiterate,[7] though one might turn this around and comment on how remarkable it is that six billion of us are literate, and how we are increasingly expressive with text. After all, reading and writing were for most of their history kept within the control, tyranny some would say, of a small elite group of religious, political and academic authorities. Literacy expanded more rapidly from the 15th century, following the invention of the printing press. Yet, even then, the intellectual resources required for developing the extended productivity of the literary brain meant that reading and writing were still the domain of an educated minority until well into the 20th century.

The *literary mind*, being one attuned to learning, developing and communicating ideas through long-form written texts, is not a timeless or universal attribute of humankind, but a particular Western phenomenon, which dominated knowledge production throughout the second half of the second millennium A.D. Nicholas Carr (2010) points out that the linear literary mind is responsible for 'the imaginative mind of the Renaissance, the rational mind of the Enlightenment, the inventive mind of the Industrial Revolution, and the subversive mind of Modernism' (p. 10). The solitary, bookish, elite and often anti-social aspects of developing the literary mind led Marshall McLuhan to claim in 1962 that print is the technology of individualism that 'detribalizes or decollectivizes man' (McLuhan 2011, p. 180). Whereas the mnemonic packaging and ritual repetitions of the oral tradition were more concerned with preservation of collectively evolved cultural wisdom and order, the literary mind is driven by new possibilities, change and the idea of 'progress'. Unlike the oral tradition where the wisdom is more important than its originators, the literary mind leads to a more ego-centric culture where originators expect attribution, honours and reward.

Building on McLuhan, Walter Ong (1982) developed the idea of secondary orality to describe radio and television, which, quite unlike the primary orality of pre-literate cultures, has a deliberate, self-conscious and permanent dependence on literate culture and texts. It is commonly claimed (e.g. Carr 2010) that McLuhan predicted that the new 'electric' media of the 20th century would lead to the 'dissolution of the linear mind' (p. 1). That prophesy seems particularly resonant with the advent of the internet, with its hypertext links which thrive on our predisposition for distraction and the gratification of quick knowledge acquisition. McLuhan's prediction also resonates with the proliferation of mobile communications, where texting and tweeting might be metaphorically described as a kind of tertiary orality. Even though the communications are not strictly oral, their interactive, collective and conversational qualities, and the phonetic shortcuts of text-speak, bear more resemblance to orality than to products of the literary mind. In many ways, new media and communication technologies are stretching and re-shaping our cognitive skills and preferences. On the one hand they make us more able to multi-task and to cope with competing sources of information. They turn many more of us into content generators, and into creators of new knowledge: we are all authors now. On the other hand they leave us in an almost continual state of overwhelmed distractedness, and (according to Carr) with less time, inclination and neural bandwidth[8] to think, understand and remember. Whether we are more or less able to differentiate critically between sources of authority and authenticity, and whether the quality of our learning and new knowledge-generation is improving or deteriorating are questions which are wide open for further research and debate.

In this context, it is difficult to conclude whether the public demand for greater author identity and authority, (and, by association copyright to protect it), is diluted or reinforced. I find Foucault's words prescient and particularly apt to a 21st-century world where meaning and relevance are ever harder to extract from the exploding babble of texts, images and sounds:

The author is therefore the ideological figure by which one marks the manner in which we fear the proliferation of meaning.

(Foucault 1969)

Fear is a key word here. How much meaning do we want, or indeed, how much can we cope with? This is a question which invites a narrative response.

Returning to Figure 14.1, which is repeated below for convenience, I would draw attention to the hope and fear plot examples at the bottom. These are part of a political repertoire where narratives of aspiration and desperation can be used as the context requires. The wiki plots should by now be recognizable as they have formed the larger part of this book, but in the context of Foucault's remark about fear of the proliferation of meaning, I wanted to emphasize the more complex distinctions between the hope and fear plots within the prevailing discourse on the right of Figure 14.1.

The hope plot has become a familiar refrain for advanced societies who feel they have a competitive advantage in the creative industries. Creativity needs nurturing because it leads to innovation which leads to economic growth, jobs, prosperity and political influence. The fear plot is more subtle. I would argue that the continuing symbolic relevance of the pirate narrative no longer primarily rests on unauthorized

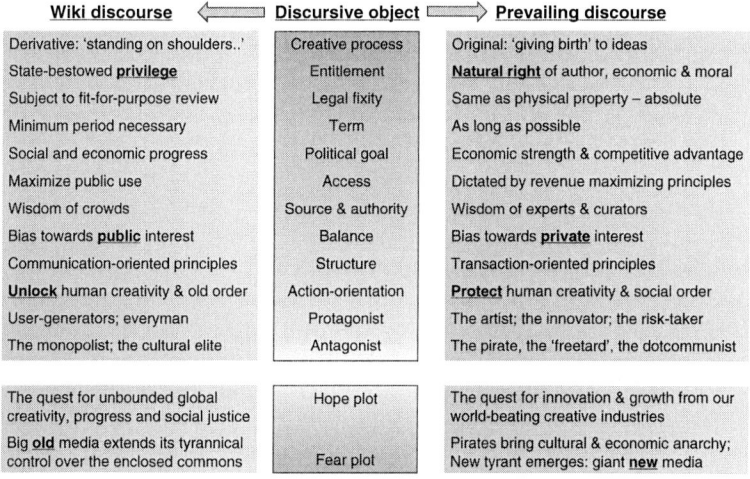

Wiki discourse	Discursive object	Prevailing discourse
Derivative: 'standing on shoulders..'	Creative process	Original: 'giving birth' to ideas
State-bestowed **privilege**	Entitlement	**Natural right** of author, economic & moral
Subject to fit-for-purpose review	Legal fixity	Same as physical property – absolute
Minimum period necessary	Term	As long as possible
Social and economic progress	Political goal	Economic strength & competitive advantage
Maximize public use	Access	Dictated by revenue maximizing principles
Wisdom of crowds	Source & authority	Wisdom of experts & curators
Bias towards **public** interest	Balance	Bias towards **private** interest
Communication-oriented principles	Structure	Transaction-oriented principles
Unlock human creativity & old order	Action-orientation	**Protect** human creativity & social order
User-generators; everyman	Protagonist	The artist; the innovator; the risk-taker
The monopolist; the cultural elite	Antagonist	The pirate, the 'freetard', the dotcommunist

The quest for unbounded global creativity, progress and social justice	Hope plot	The quest for innovation & growth from our world-beating creative industries
Big **old** media extends its tyrannical control over the enclosed commons	Fear plot	Pirates bring cultural & economic anarchy; New tyrant emerges: giant **new** media

Figure 14.1   The contested narrative of copyright

domestic downloading and casual file-sharing. That kind of 'piracy' has more or less stabilized, albeit at much higher levels than historic unauthorized copying, but still at a level where rights-owners can sustain a business from offering products and services for which those who can afford to, will actually pay. That fear of piracy has been replaced with a fear of the wisdom of crowds, and of those new players who know how to exploit the wisdom of crowds. It is a fear of losing influence and of a diminishing share in the marketplace for cultural interpretation, authority, authenticity, integrity, quality, insight and relevance. Given enough time to adapt, I have no doubt that, in the long term, big content-developing businesses could cope with much shorter and less restrictive copyrights. However, they crave more time to repackage their claim and their 'right' to be aligned with the origination process. Their larger threat of competition comes from the new giants of mass cultural distribution with whom old media companies have formed necessary but uneasy alliances. Apple, Amazon, Google and Facebook have each created products and services which are empowering and thoroughly addictive, generating huge consumer loyalty and brand power. These are so addictive that even when the shine of their public relations becomes tarnished, from time to time, with allegations of price collusion, tax-avoidance, poor labour practices, privacy invasion, and globally monopolistic ambitions for tracking public behaviour and knowledge circulation, consumers are still not sufficiently concerned with these corporations that they would boycott their products.

In this environment, the piracy narrative becomes a useful subliminal element of a more subtle fear-mongering discourse of the potential anarchy, lawlessness and corruptibility of new media. Such a discourse is sustained by those, whether old media companies or politicians, who do not want to directly attack the new giants, but believe that the traditional world of rights-owning businesses represents a necessary check and balance to their growth; growth which is, directly or indirectly, fuelled by the same disruptive phenomena which threaten and weaken the old cultural intermediaries and guardians. At a political level, this discourse even extends to the fear that these new giant corporates have become more powerful than national governments, and may become new tyrants. Government and old media therefore have a mutual interest in preserving some of the old rules and equilibrium of the cultural economy.

If the roots of US and UK intellectual property traditions *are* more public-interest-oriented than author-centric, then over the course of the 20th century that interpretation of public interest changed. It became more based upon an assumption that the extension of protection of what is mostly corporate investment in ideas, and the associated social and economic benefits of corporate products, would serve the public interest better than simply making those ideas accessible and exploitable by the general public. The economic growth of media, technology and communications companies in the 20th century may be testament to the plausibility of this assumption, but it is a patronizing assumption which is being challenged hard in the public sphere in the 21st century. *Eldred v. Ashcroft* (2003) was a hard-fought and highly publicized US legal case which challenged (unsuccessfully) the US Copyright Term Extension Act (1998). The plaintiff's argument, led by Lawrence Lessig at the Supreme Court, was that it was unconstitutional in that the retroactive extensions of copyright rendered meaningless the Constitution's statement that exclusive rights should only be granted for 'limited times'. He also argued that there had been insufficient scrutiny to ensure a fair balance between freedom of speech and the interests of copyright-holders, and that there was no empirical evidence presented to demonstrate the public benefit of transfer of public property into private ownership. *Eldred v. Ashcroft* is just one example of a fast-growing body of opinion that copyright law, as a mechanism to benefit society, is increasingly obsolete and ineffective in the digital age.

The likelihood of radical reform

Recognizing at least some of the phenomena which I have highlighted above, Government departments and institutions in the US, UK and Europe have very publicly acknowledged the need for copyright reform. I opened this chapter with Maria Pallante's recent call for 'the next great copyright act'. Similar calls have been made at regular intervals in Europe, a recent one being the autumn 2012 Lisbon Council address by Neelie Kroes, the Vice President of the European Commission responsible for the Digital Agenda. 'The legal framework has to respond'[9] she demanded. In the UK, the 2011 Hargreaves Review captures the political tone:

> we have sought never to lose sight of David Cameron's 'exam question'. Could it be true that laws designed more than three centuries ago with the express purpose of creating economic incentives for innovation by protecting creators' rights are today obstructing innovation and economic growth? The short answer is: yes. We have found that the UK's intellectual property framework, especially with regard to copyright, is falling behind what is needed.
>
> (Hargreaves 2011, foreword)

Although Hargreaves, via the British prime minister, implicitly invokes the Statute of Anne, the overarching principle of 'the encouragement of learning' has been clearly replaced by one of economic growth through innovation. Emphasizing 'the great national importance' of the British creative industries, the report claims that their digital export value 'ranks third behind only engineering and financial and professional services' (p. 3). The report is balanced in its arguments, marginally in favour of the public interest over prevailing private interests, and comes up with ten well-argued recommendations for overcoming the identified shortcomings of the current law, some of which are more concrete than others, but none which could be described as radical reform. The first recommendation echoes that of previous government commissioned reviews: that *evidence* of measurable economic benefits should drive policy rather than the *lobbying* of interested parties. Although Hargreaves does not take a position on the most sensitive topic of copyright term, the report does cite the EU extension of protection on sound recordings from 50 years to 70 years[10] as an example of policy-led decisions, rather than evidence-based policy. Another issue which is emphasized for requiring evidence-based assessment is the implementation of the Digital Economy Act's (2010) new system of law which aims to track down and sue copyright infringers, and to permit technical measures to reduce the quality, including the termination, of their internet connections. Since coming into force, the initiative to implement this key part of the Act has lost momentum, partly due to arguments over division of administrative costs between rights-holders and service providers, and due to the questionable benefits and demise of the similar 'Hadopi' system which was earlier implemented in France. It is now questionable whether it will ever be implemented.

Many of the other UK issues requiring resolution identified by Hargreaves are similar to US issues outlined in Pallante's speech, namely:

- a broader and clearer definition of limitations and exceptions to copyright, similar to the US concept of *fair use*, especially around library, archives, research, education, and data and text analytics
- legislating to release for use 'the vast treasure trove of copyright works which are effectively unavailable – "orphan works" – to which access is in practice barred because the copyright holder cannot be traced' (p. 4)
- the promotion of 'extended collective licensing' arrangements, whereby mass rights can be centrally and more efficiently administered, with rights-owners having to opt-out, rather than opt-in. The

aim of this is to give content-users the benefit of a 'one-stop shop' rather than having to spend impractical amounts of time seeking individual deals

Whilst the recommendations are far from being non-controversial (not least because many of them are seen to favour the new media giants, especially Google), neither Hargreaves nor Pallante could be accused of being radical. Nevertheless, it is worth examining Pallante's position in more detail.

As a supporter of the Stop Online Piracy Act (SOPA), which was a (thus far) unsuccessful attempt in 2011 to further expand the powers of US law enforcement to combat copyright infringement,[11] Pallante may be regarded as being conservative, albeit moderately and pragmatically, on issues of copyright reform. My first reading of her reform speech led me to believe that she was being rather timid, and it is certainly the case that someone in her position has to tip-toe through a discursive minefield, given the sensitivities of the various powerful and entrenched stakeholders. On closer analysis, however, I find seeds of a shift from the prevailing discourse towards a wiki discourse (as represented by Figure 14.1). For example, the emphasis is more on public interest and wider dissemination and usage than on issues of control, with the explicit challenge to Congress to balance the equation of 'what does and does not belong under a copyright owner's control', implying an overall diminishment of control, rather than an extension. With reference to *Eldred v. Ashcroft*, her remark that a court can only apply the facts and the law as it finds them, alongside a recommendation that Congress should re-weigh the equities of the public interest, is evidence of a certain degree of discomfort with the extra 20 years of copyright extension. An explicit suggestion that the law should shift the burden by requiring the copyright-owner to assert and register their continued interest in order to achieve those last 20 years of protection is really quite radical, even if it may seem minor and remote to most people. If a principle of copyright-holder active renewal were accepted, the conversation about shortening the minimum statutory term would become considerably less provocative.

Pallante also makes a reference to the case of *Authors Guild v. Google*, which is worthy of comment. The case involves Google's library project of mass book digitization, a process which began in 2004 and has reportedly scanned 30 million books as of 2013.[12] Although the proposal was to show only small portions of the books, presuming it to be deemed *fair use*, they scanned whole books. This was seen by the publishers as a gross breach of copyright and brought lawsuits in 2005. Many of the books scanned were still in copyright, but out of print, i.e. not regarded as commercially viable by their publishers. The case is particularly interesting because Google seemed to be attempting to address one of the historic fears and arguments of the anti-copyright lobby: that copyright can have the effect of suppressing, whether for commercial or political reasons, publicly beneficial knowledge. In objecting to the initiative, the publishers were in danger of scoring a colossal own-goal by giving a massive boost to the anti-copyright and open access movements. They therefore had to be pragmatic, and by 2008 Google and the publishers announced a settlement. It was a compromise which would let Google sell whole books and offered a subscription service for libraries and others to access the full database, with up to 20% of the book being available for free. The settlement was initially widely praised as a pioneering collaboration with far-reaching benefits for the greater dissemination and accessibility of knowledge. Perhaps unsurprisingly, it was not long before some started to object, the most ironic being Harvard University librarian Robert Darnton[13] who had been an early supporter and participant in Google's library project. The perceived problems fell into three categories. The first is a curator's concern with quality control, incompleteness or exclusion of visual materials, and integrity of scholarship, pointing out that in order to be valuable and truly accessible, digitization involves more than just scanning all available text. The second, somewhat louder, set of objections was from outside the US borders with complaints that the settlement agreement was a breach of international copyrights and treaties. The third and loudest complaint was that by creating a single access system, the project was essentially a monopolistic attempt to create the largest library and publishing business ever imagined, despite its proclaimed openness and non-exclusivity. Opponents expressing these concerns included Microsoft, Amazon and other corporate heavyweights who saw a very serious competitive threat from Google so successfully and so rapidly executing its mission to 'organize the world's information'.[14]

Despite an amended settlement agreement which tried to address some of the opposing arguments, the agreement was rejected by Judge Chin in 2011, to the relief of some, but great disappointment of others – not just the defendants in the case, but those who were, on balance, more optimistic than cynical about the public benefits of the settlement. Whilst other mass digitization initiatives are underway, such as the *Digital Public Library of America*, and the *Europeana* open culture project, these are a long way from matching the scale, pace and reach of Google's initiative.

Returning to Pallante's speech, she references Judge Chin's rejection of the Google book settlement agreement as a further example of matters more suited for Congress than the courts. This again implies that she views the outcome as being a strict application of the law, but not necessarily satisfactory from the point of view of the public interest. Along with the question mark over the outcome of *Eldred v. Ashcroft* case, and her remark that the law is essentially unintelligible to most people, her position starts to take on the appearance of being more radical than at first glance. To conclude on the topic of Pallante's speech, which I imagine may ultimately carry more weight than has thus far been acknowledged, I would draw attention to what I believe may be a deliberately provocative reference. Acknowledging the duty of Congress to keep the interests of authors in the balance, she cites (p. 26) a *New York Times* article: a 'rich culture demands contributions from authors and artists who devote thousands of hours to a work and a lifetime to their craft' (Turow et al. 2011).

The article she cites is entitled *Would the Bard Have Survived the Web?* It is a provocative polemic which takes a peculiarly one-sided approach to the debate, presuming that all reformers are by definition against copyright and against the economic incentivization of culture altogether. It argues that the structure of the Globe theatre provided a 'cultural paywall' and that its destruction in the mid-17th century by a censorious government nervous about the disruptive influence of playwrights is somehow analogous to the effect of an insufficiently policed internet upon the expression of 21st-century commercial authors. Seemingly oblivious to the 300 years of tampering, they declare that 'we tamper with [copyright] rules at our peril', 'rules that were carefully constructed by people living in the long shadow of the Dark Ages'. The vulnerabilities in the article's argument are numerous and provide fair game for ridicule. For example, Shakespeare pre-dated copyright by a century and survived very well without its protection. Indeed, if copyright had been in force, his well-known plundering of the ideas and literary material of others means that he would have been undoubtedly caught up in time-consuming and potentially ruinous lawsuits. The better question is Boyle's (2011): would Shakespeare have survived *copyright*, never mind the web? I find it difficult to imagine that Pallante's citation from this particular article is anything other than a subtle exposition of the weakness in the narrative bias in defence of the status quo.

Strategically it bears similarities to Hargreaves position (2011, p. 5) that 'the UK cannot afford to let a legal framework designed around artists impede vigorous participation' in other sectors of society.

I have chosen to focus on Hargreaves and Pallante because I find that they are the clearest political indications thus far of a meaningful shift, albeit quite small, towards the possibility of radical reform of intellectual property law. However, reform of any substance will be slow and tortuous. Pallante herself acknowledges that by referring to the fact that the still relevant 1976 US Copyright Act took two decades to negotiate, with Barbara Ringer, one of its key contributors, subsequently calling it 'a good 1950 copyright law' (Pallante 2013).

Similarly, expectations require careful management as far as the Hargreaves Review is concerned. For example, one of its key concrete recommendations was the creation of an ambitious Digital Copyright Exchange, subsequently re-labelled the Copyright Hub. The laudable idea is that the hub will serve a number of functions, including:

- information and education about copyright, helping with *navigation* and *signposting* through the maze of copyright
- a cross-sectoral registry of rights
- a marketplace for rights clearance and licensing solutions
- a source for extended collective licensing (i.e. a solution to the orphan works problem)

Although the Copyright Hub has already launched, in a very modest way, it faces some very tough challenges. The first is pragmatic. Rights are more complex than most people imagine, and it will require the establishment of technical, data and usage standards on which consensus will be difficult to achieve. Secondly it requires the collaboration of a huge number of powerful rights-holders who will be very tentative and anxious about interfacing with, and conceding to, any sort of user-focused functionality provided by an independent service. By way of illustration, at EMI I was the 'business-owner' of a project called Global Repertoire which aimed to create a similar type of hub, but just for internal use and only for the rights held by EMI. Though reasonably successful, it was fraught with negotiations with internal divisions reluctant to even slightly loosen their tight grip on clearances of rights. The project took three years to deliver at a cost of several million pounds. That was one company, in one sector, so one can only imagine the complications of building a useful and meaningful hub across not just the whole music industry,[15] but all the other rights-owning industries.

There is a third dimension to the problems of implementation of the Copyright Hub, which has some political sensitivity. Once it is in place, it is assumed that all rights-holders will voluntarily populate the database with their rights. Although not made explicit, there will then be a danger that anything not registered in the exchange will be available for exploitation. For some this looks like a story of a shift towards the public interest. For others, it is a shift towards the interest of Google, Instagram and Facebook. The main potential losers are beneficiaries of orphan works, and owners of works which are generally onerous to register, or which are not subject to secure metadata tagging, such as images and photographs. Content-owner registration may not sound like an unreasonable requirement, but it implies a fundamental shift of burden, from the prospective content-user to the content-owner, signifying a return to the practice of earlier centuries, where rights had to be actively registered rather than arising by default from the moment of expression.

A fourth and final dimension of complexity of the Copyright Hub resides in the often duplicating, and sometimes conflicting, new legislation. A current example of this is provided by Rosati (2013). Rosati questions whether the new orphan works provisions of the UK's Enterprise and Regulatory Reform Act 2013 are compatible with the recently adopted and soon to be implemented (October 2014) EU Orphan Works Directive (Directive 2012/28/EU), or with the UK's own Copyright Design and Patents Act 1988, or with earlier EU Directives and the Charter of Fundamental Rights of the European Union. She submits that 'the UK has rendered impossible any meaningful implementation of the [European] Directive into its national law' (abstract). This example is illustrative of the many esoteric legal complexities which make it extremely challenging for anyone other than specialists in copyright law to engage in a debate about copyright.

To wrap up this chapter, I would say that the current copyright reform agenda is modest and politically pragmatic. I fear that the time frame required to reach empirically defensible conclusions which are rigorous, broadly intelligible, and free of corporate and political interference is much longer than can be accommodated by changing social practice. Much of this change will happen, sooner or later, with or without the law. This means that the law is in danger of losing authority and practical enforceability as it drifts further from social norms and economic activity. Adhering to the law will therefore become (if it has not already) a tactical decision based on cost-benefit and risk analysis.

15
My Version of Events: The Future

The first three parts of this book analysed recorded music industry discourse. Part IV has been more about the discourse of knowledge production and protection, mainly because the history and the stories of 'new' technology, law and change in the much older print industry have so many insights to offer the younger recording industry. Having worked in both industries I find it difficult to separate them, especially as regards the way they are now affected by the same competing stories which will influence their future. I make no claim to be able to predict the future of cultural production any better than the next person, and I think speculating on the future is actually less valuable than analysing the past, but I will in this final chapter offer some brief views based on a distillation of my experience and intuitions. I am calling this 'my version of events' because I feel that analysis of the cultural industries will for a long time yet (and perhaps will always) lack the kind of robust empirical data on which to claim any kind of objective reliable predictions. My view is, therefore, just another tale.

Diversity and meritocracy

Trying to measure diversity in cultural output is a fruitless endeavour. As far as music is concerned, cultural critics from Theodor Adorno onwards have predicted the commoditizing and debasing effects of large-scale industrialization, and the consequential decline in diversity and quality of cultural and creative processes. Similar complaints exist in the book industry, most recently with reference to the proliferation of celebrity biographies, or books about food. Yet critics of corporate oligopoly in the music industry have failed to find convincing evidence that concentration of corporate ownership has led to any reduction in diversity. This does not surprise me, having spent the larger part of my career working for large music and publishing companies which have strived to generate business out of diversity, through the maintenance of costly and competitive internal imprint structures and extensive genre-specific development resources. Record company success is still rooted in *listening*; not just for raw musical talent, but also for cultural resonance in time and place, and with particular communities. Artists are usually already out there expressing themselves, whether in a local club or on a social-networking site, and the record company merely helps them to grow in all senses. A cultural monopoly comprising only studio-grown force-fed homogenous pop product would be as disastrous for the economics of the music industry as it would be for culture. The same goes for the book industry. David Hesmondhalgh (2013) critiques attempts to assert homogenization and to conceptualize diversity empirically or speculatively. He is of the view that the better reason to be interested in diversity is to understand 'its effects on the distribution of communicative power' and to 'reflect on whether cultural industries are providing diversity in ways that might enhance culture, society and democracy' (p. 371).

There is no doubt that the quantity of cultural products has proliferated in the 21st century. Some are concerned with whether this increases diversity, or simply multiplicity, through more copying of successful formulas and franchises, and exploiting a consumption mentality where the latest edition of a cultural product must be the best. Aside from the economic and aesthetic aspects, perhaps we should at least be happy that technology has not, as Sousa warned in 1906, destroyed the social benefits of widespread amateur expression. The relationship between the proliferation and diversity of cultural output is a topic which gained a particular public prominence with Chris Anderson's (2004) *Long Tail* article for *Wired* magazine, which formed a part of the cyber-utopian discourse of its day. Anderson argued that the discovery and recommendation algorithms of the internet, combined with the elimination of inventory-holding, and reduced fulfilment costs of e-commerce, would lead to a cultural re-balancing, diluting the dominant market-share of 'megahits' towards a longer tail of more diverse products which sell small quantities. Nielson data for US music sales in 2011 suggest that this has not happened with almost 90% of new album sales coming from only 2% (1,500 titles) of the 76,875 new titles released in the US that year. This means that the other 75 thousand titles only sold an average of less than 200 copies each. The revelation that hit-product is as popular as ever should come as no surprise, given that the mechanisms of 'word-of-mouth' or 'grapevine' now have an unprecedented global viral efficiency. Success breeds success. Faced with accessibility to overwhelming quantities and choices of cultural products, only a tiny fraction of which could be consumed in a lifetime, it is natural that people look to cultural intermediation and curation more than ever to help them filter. The bestseller list is still the easiest and most compelling source of information for most people. This of course does not necessarily equate to cultural homogenization. The actively engaged consumer has more choice than ever, and diversity may be better measured with reference to choice and availability than to actual consumption. In that sense at least, the long-tail prophesy has come true. Where this conversation gets more interesting is in the relationship between proliferation, diversity, and economic meritocracy. It has never been easier for a musician to get seen and heard, at least by a small audience, but getting paid is an altogether different matter. Every week, issues circulate the blogosphere around the question of what is *fair* in the new cultural economy. I will give just one recent example which encapsulates most of the issues.

In July 2013 the UK band Radiohead withdrew their music from the streaming service Spotify, with band members Thom Yorke and Nigel Goodrich claiming that 'new artists get paid fuck all' and that 'it's bad for new music'.[1] The outburst was newsworthy because broadly speaking, Spotify and the concept of subscription services, in general, have thus far been well received by artists and by record companies as offering a compelling alternative form of consumption and discovery of music. They have contributed to the recovery of the recorded music market and have even been seen to have a positive impact on piracy-reduction in the Scandinavian countries where their market penetration is very high.[2]

Spotify is not untypical of streaming services. The user subscription is typically £9.99 per month for unlimited access to as much music as one likes. From these proceeds, slightly less than half a penny is paid to the rights-owners each time a song is streamed by a user. That might seem very little, but it is payable every time a song is played, and the user does not own the music; if the user stops paying the subscription it is no longer available. In 2012, songs by the UK artist Adele were streamed 163 million times via various streaming services in the US alone.[3] Clearly this is the high end of activity. New artists who are listened to by only a few hundred people are scarcely going to notice a Spotify contribution on their royalty statements. In percentage terms, the ratios between what the consumer pays, what the retailer (Spotify) keeps, and what gets paid through to the rights-owners (to be divided between record company and artists) are broadly the same as the 'old world' of music purchases to own. So in this sense the business supply economics have not really changed. What the subscription model does do, however, is to change the economics of consumption, promoting music discovery and removing most of the risk of consumer purchasing. For active music consumers, i.e. those who were previously buying, say, five albums a month (either CDs or iTunes downloads), Spotify represents exceptional value for money. This value is not only in the cheaper price, which is in the region of 70% less than purchasing to own, but also lies in the opportunity to explore and listen to many more products than in the purchase-to-own model. No expensive mistakes from buying over-hyped albums, and no financial obstacles for experimenting with unknown artists. In this sense, the model seems to promote both diversity and meritocracy. However, one could argue that if Spotify only ever comprises highly active music consumers then the artists will lose out versus the old system as they will have borne the cost of the cheaper price. This may just be a matter of timing. Active music buyers traditionally made up less than 20% of the listening population. If Spotify, along with rival subscription services, can achieve the status of being an essential utility and reach the same kind of market penetration as that enjoyed by pay TV services (i.e. the majority of UK households) then the £9.99 paid by millions of people who previously spent much less on music will represent a real incremental expansion in the music market. The same may be eventually true of the consumer book market, though that seems more remote at the time of writing.

Going deeper into the competing interpretations of 'all-you-can-eat' subscription services and their impact on cultural production, what critics may be really lamenting is the debased experience of listening. In the old days, buying an album was a real commitment. You had to do your homework before deciding what to buy, and you had to save your pocket money to make that one special purchase. When you got the record or CD home you played it intensively, even if at first listening it was not as good as you had imagined it would be. By contrast, browsing on a subscription service is 'too easy'. There is no commitment. It may be addictive, but one can be very easily distracted by 20 million songs, favouring music which instantly gratifies over more worthy but less instantly digestible work. There is a risk that the music consumption experience becomes more superficial.

Personally, I do not agree with such a view, and I wonder if what is really going on with Radiohead's Spotify outburst is a gesture of guilt-tinged solidarity by the older artists, in recognition that the new artists may never have it so good as under the old pre-internet, high-priced portfolio model of cultural production. There, the last vestige of a centuries old culture of patronage accommodated greater security and funding of artists through the early nurturing period, and distributed super-profits from hit artists to absorb losses on the majority of new ones. With fewer barriers to entry for new artists, and greater price elasticity amongst consumers who have more consumption choices, the resulting lower proceeds from recorded music are now spread more thinly over a proliferation of new artists. The new diluted meritocracy is perhaps just too brutally fair for those who are concerned about a diminishing profile in society for the aspiring 'artist', as opposed to the jobbing musician.

The demise of the portfolio model

Cultural production will always depend upon a large pool of amateur and part-time professional activity from which to cream off the best talent. This of course leads to an understandably cynical view that record companies waste more talent than they nurture. It was argued by Miege (1989), amongst others, that the permanent oversupply of creative workers means that they bear the unrecognized costs of conception by being willing to forego the benefits of secure working conditions. This phenomenon tends to be obscured by the fact that very generous rewards are available for 'star' talent who achieve public recognition and a loyal following. The greater negotiating power afforded by a star's scarcity value means that a greater proportion of the risks and costs are borne by producing companies. Star culture in industry can be traced back more than half a century and has shaped the economic portfolio model. Much of the economic justification of the product pricing and division of rewards in the cultural industries has been based on the portfolio approach. This means that most new talent received a level of investment which allowed them a chance of stardom. But as star creation is not a precise science, the 'hit-rate' for stars is low. As a result, companies lose money on the majority of new investments. High margins are thus required on the minority which are successful in order to finance continuing high-risk investment in new talent. The degree to which this portfolio model represents a just and meritocratic process based on universal natural or human phenomena, as opposed to being heavily subject to inequitable control by the powerful and wealthy, has always been highly contested. The model might now become obsolete and give way to an alternative meritocracy where lower product pricing disincentivizes producing companies to take such big, or so many, risks and puts a correspondingly higher burden on talent to support itself through its own development. This may be over-theorizing, but either way, there is consensus that new creators have to work harder, not only to understand and take advantage of the new business models, but also because their position is more precarious in the new meritocracy. This is as true for authors as it is for musicians, though for now at least, the book is holding its price-value in the eyes of consumers. Perhaps this is because long-form 'linear' books remain a rarer type of thing, a strangely contemplative sanctuary from the hyper-texting distractions of the shallows.

User-generated utopia or data-dominated dystopia?

Diversity and economic meritocracy aside, the narratives are still raging about whether the proliferation of cultural texts, in the widest sense, will make the world more transparent, stable and safe. A couple of years ago I attended a dinner at my then corporate neighbour, *The Guardian* newspaper. The conversation was still abuzz with the Arab Spring; in particular the role played by mobile phones, Twitter and Facebook in helping 'the people' mobilize, and in particular the way *The Guardian* had so successfully plugged into an informal network of media-savvy citizens. Aside from the usual cynical and envious grumbles that *The Guardian* was fortunate enough to invest so much in new technology only because it did not have to worry about figuring out a profitable business model, there was a constructive curiosity about the changing roles, identities, information sources and interactive relationships amongst news and content consumers and generators in the digital age. The conversations were just one small example of a maturing discourse which began decades ago with the construct of the computer as a means of liberation, and now extends to the broader emancipatory effects of new media and technology in the 21st century, and a persistent optimism that a fairer, more open and transparent world is possible through technology. Anyone can now get his or her voice, image, music, photos, videos, opinions, secrets, lies, fantasies, revelations, insights and news 'out there' on an equal footing.

As we get further into the second decade of the new millennium, the technological utopia has all but evaporated along with the peaceful democracies imagined by the Arab Spring. Evgeny Morozov's *The Net Delusion* (2011) provides a scathing critique of those who claim that the internet is an inherently democratizing phenomenon, and gives ample illustration of cultural, political and economic repression by governments around the world, aided and abetted by the internet. Eran Fisher (2010) similarly pricks the discursive bubble of networked technology as a vehicle for social emancipation, demonstrating instead how it legitimates a new form of capitalism. Not all books are as bleak and apocalyptic as Al Gore's *The Future* (2013), but seemingly no book about the future can now be taken seriously unless it has a dark and sinister edge.

On the theme of data ownership and the socio-economic justice of the internet, even the irrepressibly optimistic Jaron Lanier, writer, musician, computer scientist and pioneer of virtual reality, finds dark forces within the 'siren servers' in his book *Who Owns the Future?* (Lanier 2013). A siren server is an elite computer or network which provides services which we cannot resist and which gathers data without paying for it. It is 'characterized by narcissism, hyper-amplified risk aversion, and extreme information asymmetry. It is the winner of an all-or-nothing contest', secretively analysing Big Data and using the results to 'manipulate the rest of the world to advantage' (pp. 49–50). Being an optimist, Lanier comes up with an extraordinary and ambitious solution to redress the tyranny of the siren servers over internet users. In doing so he implicitly accepts the inevitability that we are living in an increasingly post-private world. In exchange for giving up our privacy and our lucrative behavioural data, our economic dignity would be restored by being compensated a tiny royalty every time our data is commercially used. So, for example, if you use a dating site, whether successfully or unsuccessfully, and your profile data is used algorithmically to help to match another couple, even years later, you would receive a nano-payment. Aggregated across all media-tracked activities, and over time, this could add up to a meaningful number and represent a fairer burden of cost for the siren servers. The idea is visionary and is a helpful new articulation of some tricky problems, but it rests on the assumption that the gap between technical possibility and its execution by flawed individuals, organizations and institutions can be bridged. Given the music industry's experience with the complexity of digital rights databases and royalty systems, and the modest timeline projections of the Copyright Hub, I do not think we can hold our breath waiting for Lanier's solution. Nevertheless, his moderation of the discourse of technological determinism in favour of the internet user helps to address Reyman's (2010) complaint that 'the role of internet user as creator is often absent, or at best lacks agency in the debate' (p. 7).

Final word

It is difficult to stand back from cultural production and assess what has really changed and what remains the same. I have tried to tread a path between the transformational hype and revolutionary sensationalism of some, and the rigid conservatism of others. I will conclude by trying to address the 'so what?' question – what's my point? Is there a problem which needs solving here? And what have I added to these entrenched cultural-economic issues which are already widely and deeply studied and written about? In brief, it is this: to try to diffuse some tension in the long war between private capital protection of cultural output and the public interest by making the narrative *agon* or conflict more transparent; to illustrate that copyright is not 'carefully constructed' in the shadow of the Dark Ages, as was dubiously asserted in Chapter 14; that it is not fragile; and that we should not treat it as being so. It is socially and rather messily constructed from centuries of promoting competing stories in order to make sense of commonly experienced social dilemmas. Similarly, the processes of cultural and knowledge production are not fragile. They are deeply embedded qualities of human social behaviour and pre-date copyright. They will survive any 'tampering' with intellectual property law. The storytelling contest itself is simply a metaphor designed to contribute a critical perspective to the important negotiations around customs, practices and laws which require modification from time to time to reflect changing economic, cultural, political and technological circumstances. The balance to be struck is between freedom of expression, social stability, social justice, cultural aspiration, cultural heritage, technological innovation, commercially driven prosperity, and international competitive advantage. In my experience, much of the antagonism between opposing parties is not actually based on ideological differences about where these balances should be struck, but comes more from distrust of personal agendas and concerns over the technical and executive competencies of policy-makers.

Why is it important to make these statements? Because the stories are shifting more quickly than can be accommodated by the law and the policy-makers, and people are getting very animated about it. The continuing proliferation of digitally available content will increasingly spill over the walls of artificial scarcity. The new stories are not particularly new. The narrative landscapes may be contemporary, and some of the characters have changed their names, but the new balances to be struck are between stakeholders whose plotlines would have been recognizable in the 18th and 19th centuries. Historical analysis can help. The political risk-appetite for any radical shake-up of intellectual property seems small, and this is partly because the protagonist of the user-creator is still a rather colourless character in a more culturally democratic *wiki* narrative. Rights-owners will therefore likely hold onto the current protections for the foreseeable future, but in the much longer term, the 21st- and 22nd-century survivors will be those players who are focused on creating new oligarchies and monopolies. These will be based not on copyright, but on building organizational strength from innovation, or on being sensitive to emerging culture, and in either case having an unwavering focus on products and services which customers, be they users or generators, actually value. I therefore believe that the longer the legacy content-holders devote extensive time and resources to their medium and long-term capital protection strategies, the less likely it is that they will hold their ground in the competition for expert services and products.

I have endeavoured to remain relatively politically neutral on the topic of copyright reform because my main interest is in demonstrating the dynamics of language and stories rather than the identification of an optimum political outcome. However, if the reader wants a less equivocal expression of my position, then I will say that I am not an active advocate for *radical* reform of copyright, nor for its demise. I am not opposed to reform in principle, but I do not think that copyright is yet perceived as a big enough problem to demand a radical resolution, and it would be presumptuous of me to help construct it so. Radical reform will, and should, happen if reasonable new ideas gain enough plausibility to motivate enough people, and if those ideas can be executed with a meaningful outcome. Ideas of flat-fee compulsory licensing, or Fisher's (2004) proposal to replace copyright with a government-administered reward system involving free access for consumers, funded through general taxation, have some merit, but their momentum has waned. By contrast, Creative Commons is a good example of new copyright practice which I hope will continue to change the way copyright is perceived, in particular how 'all rights reserved' is a lazy option which is over-used to the detriment of society.

Intellectually I can see a very strong argument in favour of maintaining copyright, but with clearer 'fair use' exceptions and a much shorter term, with conditions for renewability. Any extensions to copyright term, such as the recently implemented European Directive, extending the protection of sound recordings from 50 years to 70 years, feel counter-intuitive and may be a Pyrrhic victory for the industry – adding little in economic value, but needlessly antagonizing the anti-copyright lobby and giving them fuel for their arguments. I do not believe there is a magic number of years of protection. Such a number would never be proven empirically to the satisfaction of private and public interests, but personally I do not see a clear argument for initial copyright term being longer than for patents, especially if it were renewable. Fierce opposition from existing content-holders, along with the enormous practical and political difficulties of getting international agreement, makes it seem very unlikely that a radically reduced copyright term will be on any government's agenda in the near term.

Nevertheless, I do believe that life plus 70 years will be increasingly seen as bizarre and untenable to younger generations, who are already accustomed to broader access to cultural products than previous generations, and also accustomed to creating and sharing things without automatically expecting payment. Open Access is already well advanced in its seemingly inexorable journey to become the dominant business model for scientific publishing, and the forward-thinking publishers have understood and acted on their obligation to redefine their contribution to the value chain. Educational publishing may follow in the same route, whether voluntarily or through political coercion, further diluting the domain of copyright and its power as a social construct to younger minds. So whilst the desire to make a living from the creation of content is an aspiration which is more common than ever amongst the young, expecting a pension based on copyright will, I believe, be less recognizable and less intuitive for them.

In any event, the creative industries are sufficiently robust to withstand copyright reform. Their core strength, value and longevity lie more in their inimitable and intangible processes, networks, and people, than in the length and maximum exclusivity of their copyrights. After all, the recorded music industry was blindsided by one of the most sudden and dramatic industrial disruptions of modern times. Within three years, consumer acquisition[4] of recorded music moved from predominantly authorized, to predominantly unauthorized. Yet the industry has adjusted to a smaller market for *recorded* music and survived with margin percentages not much below where they were in the booming 1990s. Although it has become harder for artists to get rich out of making music (it was never easy), the opportunities for musicians to at least be heard, and to build a modest income from their efforts, have arguably never been better, especially for those who understand exactly what it is they have to offer, and can take advantage of the different economic mechanisms open to them.

Far from destroying the creative industries, laws which could resonate more intuitively with the creative and social practices of the 21st century could strengthen public respect and compliance, and make enforcement more focused and effective. Good compliance with a minimal copyright regime would be economically more attractive than the flagrant abuse and frustrations associated with a maximal regime. A naïve and idealistic view? Maybe, and obviously any extreme overnight change would wreak havoc on corporate balance sheets and budgets. It would also bring real hardship to individual creators, especially to older ones who have become dependent on royalty income streams. But the debate should be about the next 100 years, not the next five or ten, and well-designed transitional measures, planned and applied over a long enough period, could eliminate much of this disruption.

The conflict between piracy and policing will be around for a long time yet, but it is a battle which will never be won. The outcome may always have to be a managed equilibrium where content-users make a risk assessment, consciously or unconsciously, whether or not to infringe, and content-owners make a cost-benefit analysis of how much to spend on anti-piracy strategies. This is where we are today, but it is arguably not socially optimal in societies which claim to be progressive. The properties of new technology and communications do offer viable alternatives to exclusive content-ownership but they need to be negotiated, both by experts and through public discourse. There is a point where arguments about copyright become exhausted, or at least exhausting. The study of intellectual property law and social policy is fascinating and the field is populated with some exceptionally talented people. It is also overwhelmingly complex and esoteric, severely weakening its priority on the public agenda and the scope of its political impact. A high turnover in UK government intellectual property ministers may be a testament to that. But to those who are directly or indirectly involved in the pursuit of a more manageable, better understood and respected equilibrium in intellectual property law, I would simply say: don't give up, it is a worthy quest.

Notes

Introduction: A Changing Master-Narrative of Cultural Production

1. Although this quote has been popularly and plausibly attributed to Plato, the attribution appears to be unreliable. A more reliable source of a similar sentiment is the Scottish nationalist Andrew Fletcher (1655–1716) who wrote: 'if a man were permitted to make all the ballads, he need not care who should make the laws of a nation'. I have deliberately left the attribution to Plato in order to illustrate the question of authority and social power of plausibility which are themes of this book.

1 A Personal Perspective

1. *The 25 Worst Tech Products of All Time*. PC World, 26th May 2006. www. pcworld.com

2 Innovation or Bust: A Short History of Recorded Music

1. The amended painting was commissioned by William Barry Owen, head of The Gramophone Company, the British company formed to exploit Berliner's European patent rights.
2. According to the US Census figures cited in Gelatt (1955, p. 143).
3. Just who coined this phrase has never quite been determined. An English electrical engineer, Harold A Hartley, is the strongest contender; he claims to have invented the phrase about the end of 1926. [...] If this be so, 'high-fidelity' lay fallow for some time. Not until late 1933 or early 1934 did the phrase come into general use, but then it was exploited with a vengeance.

   (Gelatt 1955, p. 207)

4. The practice of record companies giving incentives to DJs to promote their records has been the subject of repeated state investigation over the decades, the most notable being the 'payola' scandal of the late 1950s and early 1960s involving the DJ Alan Freed.
5. The actual peak year varies depending on whether the US, UK or global market is measured, and depending on whether units or value are considered. For current purposes, these details would be a distraction from the overall narrative. IFPI total global unit data for all formats indicate that 1981 was

the peak year with almost 2.2 billion units, but the US and UK peaked earlier than this.

6. The anecdotes centred on PolyGram's colourful but financially disastrous relationship with the infamous Casablanca Records and its head, Neil Bogart. Casablanca was the label responsible for the success of Donna Summer and the Village People, aided by extravagant business practices which fuelled what Mark Coleman (2005) calls the 'Roman candle flameout' (p. 130) not only of Casablanca Records, but the disco movement itself.

7. There is a counter-argument that home-taping and sharing resulted in the promotional dissemination and wider discovery of music at a time when radio channels were quite limited in their coverage of new music. Whilst some sales are lost, a wider audience and new fans are gained virally. Also, a market with multiple formats (LP/MC/CD) leads to greater consumption than a single-format market, as more affluent consumers buy the same product in different formats. Whilst there has been much rhetoric on the home-taping topic, there is, to my knowledge, no definitive data which determine the net impact of these opposing phenomena.

8. On its 20th anniversary in 1999, Sony had sold 186 million units of the cassette Walkman. Other brands, notably Panasonic, also had huge success with portable cassette players on the back of Sony's pioneering success.

9. An internet search on 'vinyl versus CD' illustrates the long and unresolved debate about whether CD is higher quality than vinyl. Whilst some countries have seen spectacular growth (in % terms) in vinyl records in recent years, and vinyl forms a significant element of the initial format offering for some artists and genres, it remains a niche format.

10. The European Competition Commission approved Universal's acquisition of EMI subject to conditions including the disposal of certain assets, notably the Parlophone label which was subsequently acquired by Warner Music.

11. With hindsight, Philips had a lucrative exit from its 75% PolyGram holding. By contrast, Bronfman is often criticized for losing US$ billions of his family's shareholder value through his entertainment industry transactions.

12. In a classic case of closing the stable door after the horse has bolted, the industry experimented with copy-protected CDs. After much public criticism, culminating in the 2005 outrage at Sony-BMG's products which rendered Windows PCs vulnerable to malware, copy protection was effectively abandoned in recognition of the fact that it would only accelerate the demise of the CD format.

13. Nowhere is this better illustrated than by the proposed music download service Pono (www.ponomusic.com), established by the musician Neil Young, whose aims are to present songs as they first sounded during studio recording sessions using high-resolution audio instead of the compressed audio inferiority that MP3s offer. Pono is a Hawaiian word meaning 'righteous'.

14. MPEG was established by Italian Leonardo Chiariglione with Japan's Hiroshi Yasuda, and the early network involved scientists and technologists from Germany, the US, the UK, Canada, France and the Netherlands.

15. Michael Kuhn (2002) thus describes his PolyGram management colleagues, with affection and respect.

16. It was concentrating on electronic hardware products which it now describes as having 'a focus on people's health and well-being', Philips 2012 annual report: www.philips.com

17. Source: apple.com/pr/library
18. About 20% of the 350 million iPods sold were the iTouch models, which arguably were not simply bought as music devices. However, I have included no revenues from iPhones and iPads, which have certainly benefited to some extent from customers who feel committed, or tied, to the Apple platform through their music collections.
19. Nielsen Soundscan reported a decline in US digital track sales ('Digital Music Sales Decrease For First Time in 2013' billboard.co.m/biz, 3rd January 2014.
20. 'Pandora Announces May 2014 Audience Metrics' www.investor. pandora.com
21. According to some analytical sources, Apple has a 75% market share ('Apple's iTunes accounts for 75% of global digital music market, worth $6.9B a year.' appleinsider.com., 20th June 2013).
22. A popular analogy of music being supplied as a commoditized public utility, like water or electricity, was coined by David Bowie in a *New York Times* interview (9th June 2002, nytimes.com). This same concept of 'music like water' was central to Kusek and Leonhard's well-known polemic book, *The Future of Music: Manifesto for a Digital Music Revolution* (2005).
23. I will consider the dilutive effects of subscription models in Chapter 15.
24. This anti-subscription sentiment was clearly expressed in Steve Jobs' interview with Rolling Stone magazine, 3rd December 2003.
25. Apple CEO Tim Cook, quoted in the Financial Times. ft.com, 29th May 2014.
26. 'Apple to reach 600 million users by end of 2013' cnet.com., 4th June 2013. Note that this refers to users of all Apple products, not just music products.
27. According to Wikipedia (entry under Beats Electronics), Iovine recalls Dr. Dre saying to him: 'Man, it's one thing that people steal my music. It's another thing to destroy the feeling of what I've worked on.'
28. According to Spotify's website: press.spotify.com
29. Calculating market shares in the new millennium is notoriously difficult given the multiple sources of revenues. This approximate illustrative calculation is based on Spotify's gross revenues assuming ten million users, and iTunes payments to record companies, reported on cultofmac.com (27th February 2013), grossed up by 10/7 to make them comparable with Spotify's sales.
30. Sony Music and BMG merged through a 50:50 joint venture in 2004. An EU competition commission investigation into the merger spoilt the chances for a long-desired similar merger between EMI and Warner Music in 2006. EMI was subsequently acquired by Universal in 2012, though the EU required Universal to divest some of the EMI assets. Consequently EMI's best-known label (Parlophone) was acquired by Warner Music in 2013. SonyBMG ceased to exist when Sony bought out Bertelsmann's share of the joint venture in 2008, re-branding as Sony Music Entertainment.
31. 2012 data showed the first year of global market increase (0.2%) since 1999, suggesting that the decline may be at an end and that the recording industry may have re-stabilized at this lower level. The market did fall by a further 3.9% in 2013, but this was heavily influenced by a 17% fall in Japan, which if excluded from the data shows that the rest of the market was flat, with modest growth in the US and Western Europe offset by notable falls in Australia and Spain. Source: www.ifpi.org

32. Data from the UK collecting society PRS show that whilst recorded music consumption as a percentage of total consumer wallet-share has dropped by more than half since 1999, live music increased more than fourfold by the same measure. In the ten years from 1999 to 2009, aggregate music consumption has therefore grown in absolute terms, and lost only 15% of its value measured by share of the total consumer wallet. prsformusic.com/creators/news/research/documents/economic%20insight

3   Value Shift

1. Job titles were applicable in 2008.
2. Cumulative sales of iPods are 350 million units as of September 2012, according to Apple's website.
3. Study from the University of Hertfordshire reported in *The Times* (London) 16 June 2008.
4. Data from *Wired* magazine, 31 July 2008.
5. *New Musical Express* – 24 June 2008.
6. Comments from Page & Garland, reported in *Wired* magazine, 31 July 2008.
7. Source: PRS For Music *Economic Insight*, 18 April 2011.
8. En.wikipedia.org/wiki/List_of_highest-grossing_concert_tours, 14 June 2013. Wikipedia sources are incomplete, but the chart is reliable enough to support the claim, without controversy, that growth in the live music market has been driven by mature and established acts.
9. Apple CEO Tim Cook, quoted in the Financial Times. ft.com, 29th May 2014.
10. It was briefly valued at $12 billion when News Corporation attempted to merge it with Yahoo in 2007 (*The Daily Telegraph*, 24 March 2011. www.telegraph.co.uk)
11. *Bloomberg Business Week,* 29 June 2011. www.businessweek.com

4   Custodial Tensions

1. EMI commissioned research, also consistent with IFPI piracy data.
2. Except for the period 2004–2006 when some major record companies experimented unsuccessfully with 'copy-control' technologies which prevented the loading of CD content onto computers (mainly outside the UK and the US markets).
3. The Copyright Alert System (CAS) known as the 'six-strikes rule' was implemented in the US in February 2013.
4. BBC Radio 4. *Today* programme. 24th July 2008.

5   Hindsight and Strategic Sense-Making

1. It eventually settled lawsuits brought by Universal and EMI for an amount reported to be in excess of $US 100 million, and funded by the deep pockets of Bertelsmann, who had made a strategic acquisition of Napster before it was forced to close.

2. It was leaked that the company had a £200,000 head office bill for 'the now legendary *fruit and flowers'*. This was enthusiastically interpreted by the press as being 'a music biz accounting euphemism for the provision of drugs' and other 'artist services' (*The Guardian*).

7 Strategy as Storytelling

1. This was the last of many attempts by EMI and Warner to merge over the previous decade. Warner eventually picked up EMI's Parlophone label in 2013 after partial disposals were made by Universal Music Group in order to satisfy EU commission approval requirements for Universal's acquisition of EMI.

10 The Inventor's Tale

1. One notable omission from Figure 10.1 is the greater accessibility and affordability of domestic recording software, but because it is a relatively small specialized market, innovators and exploiting companies do not attract much attention.
2. After losing its legal battle, Napster subsequently pursued authorized commercial opportunities, though mostly unsuccessfully. Fanning had some success with derivative ideas such as Snocap, a legalized file-sharing solution, which was sold to Imeem in 2008. Imeem was sold the following year in a fire sale to MySpace.
3. *shashwatdc.blogspot.co.uk/2007/09/interview-karlheinz-brandenburg.html*
4. PPL is the UK central licensor for recorded music use in broadcast, public performance and new media.
5. PRS is the UK body which licenses musical compositions, i.e. it represents the songwriters rather than the recording artists.
6. Logic Pro provides software instruments, synthesizers, audio effects and facilities for computer recording and mixing. It was not invented by Apple, but acquired by Apple in 2002. Pro Tool is a competing digital audio workstation for Microsoft Windows.
7. A popular tale is that the Beatles were responsible for the invention of the CT scanner which was invented by EMI in the early 1970s, the research laboratories (CRL) being financially supported by EMI's music division.
8. Eliot Spitzer was the New York State Attorney General and well-known prosecutor of white-collar crime. He pursued the record industry for many years through investigations of payola and price-fixing. To the amusement of many industry executives, he resigned in March 2008 in the wake of his involvement in a high-priced prostitution ring.

11 Power and Ideology

1. The UK government Department of Culture Media and Sport.
2. A rivalrous and non-durable good (such as a pie) means that its enjoyment by one consumer prevents anyone else from enjoying it. By contrast, in the age

of recorded technology, music has become a non-rivalrous good: its consumption by one person does not constrain or diminish its enjoyment by others. Exclusive or sell-out concerts, and rare and undistributed recordings could be argued as exceptions to this generalization.

3. Inscriptions included: 'not licensed for radio broadcast', and 'licensed for non-commercial use on phonographs in homes.'

4. As of 2013 (the first full year after the break-up of EMI), three companies will represent around 75% of the global market for recorded music.

12 Pirates, Property and Privatization

1. The logo was reincarnated with some irony in the new millennium within the mainsail of the square-rigged pirate ship logo of the file-sharing facilitator, The Pirate Bay.

2. Appeal by Phillip Michael Jackson, 13 November 2012. guardian.co.uk.

3. Case of Melchior Reitveldt versus Buma/Stemra. Cited in *Wired* magazine, 18 July 2012.

4. A proposition made by Twenge, J. (2006) *Generation Me*. New York. Free Press, which has gained notable currency in demographic summaries and marketing assumptions.

5. This uncontroversial view is attributed by Finlo Rohrer of BBC Magazine (18 June 2009) to Dr. Richard Jones, author of Entertaining Code: File Sharing, Digital Rights Management Regimes, and Criminological Theories of Compliance. International Review of Law, Computers and Technology 2005, Volume 19, Issue 3.

6. Thomas Hobbes refers to shooting at one another with paper. *Behemoth* 1668. (Google eBook, location 1970).

7. The full name of the 1662 statute was *An Act for preventing the frequent Abuses in printing seditious treasonable and unlicensed Bookes and Pamphlets and for regulating of Printing and Printing Presses*. Source: *Statutes of the Realm*; volume 5, 1628–1680. Editor: John Raithby. History of Parliament Trust. www.british-history.ac.uk

8. Atkyns' story went that Henry VI sent Robert Turnour with William Caxton to Flanders to bribe one of Gutenberg's workers (Frederick Corsellis) to come to Oxford to set up printing in England, with his first book appearing in 1468, predating Caxton's own claim to have printed the first book in England in 1471. This story is called into question in Philip Luckombe's (1771) book, The History and Art of Printing.

9. Guinness Book of World Records archive 27 November 2005.

10. My references for such accounts include: Johnson (1998); Parker (2009); Pirate Utopias (anon) published in Do or Die (Issue 8, 1999); Burl (2006).

11. Johnson lists 11 components of such articles, attributed to Captain Roberts, which even include compensation for injury in service.

12. Interestingly, the Project Gutenberg eBook edition of volume 2 of *The General History..*, which is from the Henry E. Huntington Library's first edition copy, is simply called The History of the Pyrates Vol. II, and lists Daniel Defoe as the other, despite considerable academic disagreement about Defoe's authorship.

13. Parker points out that Errol Flynn destroyed Libertatia (an alternative name for Libertalia) in order to keep the ships of the East India Company safe, in the 1952 film *Against All Flags*.
14. I have cited this Project Gutenberg eBook source to illustrate the confidence with which some scholars/editors attribute *The General History* to Defoe.
15. In German elections in 2011, the Pirate Party gained around 8% of the vote in many regions, according to Spiegel Online International. www.spiegel. de/international/germany/pirate-party. In Sweden in 2009, the Pirate Party achieved 7% of the vote in European elections. www.theguardian.com. However, the popularity of various national pirate parties seems to have fallen since these high points.

13   Enclosing the Commons of the Mind

1. Hardin acknowledges that we do not know that the commons is not infinite, but counters that 'we will greatly increase human misery if we do not, during the immediate future, assume that the world available to the terrestrial human population is finite' (p. 1243).
2. The global birth rate may have stabilized in the last decade, but the ageing of today's higher number of children, combined with the increased longevity of the total population mean that total population will continue to increase to ten billion at least (UN projections). There is some consensus that total population will peak during the 21st century, though there is wide variance on precisely when and at what total number.
3. Boyle (2008, p. 9) estimates that as much as 95% of the Library of Congress catalogue is commercially unavailable. He also point out that when the US copyright required renewal after 28 years, about 85% of copyright-holders did not bother to renew.
4. According to their website (creative commons.org), an estimated 350 million CC licences had been used as of 2009.
5. A Declaration of the Independence of Cyberspace. John Perry Barlow's address at Davos 1996. Projects.eff.org/~barlow/declaration-final.html
6. Urbandictionary.com
7. www.ted.com/talks/lawrence_lessig_we_the_people_and_the_republic

14   The 300-Year War of Copyright

1. Transcript of a lecture delivered at Columbia University, 4th March 2013. web.law.columbia.edu/sites/default/files/microsites/kernochan/files/Pallante
2. Mark Rose (2010, p. 70) interprets Habermas' concept of hollowing out of the public sphere as a process beginning in the 19th century where political figures and large organizations manipulate mass media for commercial interests, and the state assumes regulatory and protective functions in civil society, all in new feudal ritual of authority.
3. Also popularly known as the Sonny Bono Act, or the Mickey Mouse Protection Act. It also extended works for hire, anonymous and pseudonymous works to the shorter of 95 years from publication or 120 years from creation.

4. Attributed by John of Salisbury to Bernard of Chartres. Source: Wikipedia, citing *The Metalogicon of John Salisbury*. University of California, p. 167.
5. The passage is an excerpt from a letter to Isaac McPherson in response to his letter to Jefferson on the question of the expiration of a patent for 'Elevators, Conveyors, and Hopper-boys'.
6. From Benjamin Jowett's translation of Plato's Phaedrus. A Public Domain Book (Kindle edition).
7. 84.1% of the world's population is literate, according to Wikipedia 23rd August 2013, based on data from the CIA World Factbook and national self-reported data.
8. Carr (2010) cites a number of scientific sources to demonstrate that heavy internet users have diminished capacity for contemplation, reflection, abstraction, recall, critical thinking and imagination. The 'outsourcing' of memory to the web has consequences we do not yet fully understand.
9. Europa.eu/rapid/press-release_SPEECH-12-592_en.htm
10. Sound recordings and performers' rights had protection of 50 years from recording date in the UK. A 2011 EU harmonizing Directive requires an extension to 70 years to be implemented by 1st November 2013. Source: UK Intellectual Property Office. www.ipo.gov.uk
11. SOPA was opposed by various web communities, with the help of Wikipedia and Google, who themselves drew criticism for wielding influence on a topic in which they were perceived to have a clear interest.
12. According to Robert Darnton in *The New York Review of Books*, 25th April 2013. www.nybooks.com.
   In 2010 Google made an estimate of the total number of books in the world and arrived at a number of around 130 million, though the definition of a book is questionable, and the process of approximation is arguably rather meaningless.
13. Darnton is an advocate of the Digital Public Library of America, a service which addresses some of the issues that were seen to be problematic with Google's approach.
14. www.google.com/about/company/
15. An initiative for the creation of a global music publishing rights database is already underway. www.globalrepertoiredatabase.com.

15   My Version of Events: The Future

1. Comments from Radiohead band members Thom Yorke and Nigel Godrich, as reported on www.theguardian.com, 15th July 2013. In a more recent outburst, Radiohead's Ed O'Brien, as co-chair of the musicians rights campaign group, the Featured Artists Coalition, has criticized YouTube for undermining artists even more than Spotify in its proposed new subscription service, claiming that they 'risk creating an internet just for the superstars and big business' bbc.co.uk/news/technology, 4th June 2014.
2. Research is still relatively undeveloped, and subject to bias, but there have been a number of studies which suggest that the growth in usage of good legal

business models (including subscription models) reduces unauthorized usage. Recent example can be viewed at: techdirt.com/blog/innovation/articles/20130723/12235723906/two-new-reports-confirm-best-way-to-reduce-piracy-dramatically-is-to-offer-good-legal-alternatives.shtml

3. According to Nielsen's 2012 Music Industry Report www.businesswire.com

4. I say acquisition rather than enjoyment, because the quantities of music downloaded and stored on computers and portable devices well exceed levels of what could be reasonably listened to by the average consumer.

References

Anderson, C. (2004). 'The Long Tail.' www.wired.com/wired/archive/12.10/tail. html.

Andrews, K. R. (1971). *The Concept of Corporate Strategy*. Homewood, Illinois, Irwin.

Ansoff, H. I. (1965). *Corporate Strategy*. New York, McGraw-Hill.

Atkyns, R. (1664). *The Original and Growth of Printing, Collected Out of History and the Records of This Kingdom: Wherein Is also Demonstrated That Printing Appertaineth to the Prerogative Royal and Is a Flower of the Crown of England*. London, cited by Luckombe 1771.

Balogun, J., Anne Sigismund Huff, Phyl Johnson (2003). 'Three responses to the methodological challenges of studying strategizing.' *Journal of Management Studies* **40** (1): 197–221.

Barry, D. and M. Elmes (1997). 'Strategy retold: Toward a narrative view of strategic discourse.' *Academy of Management Review* **22** (2): 429–452.

BERR and DCMS (2008). Creative Britain: New Talents for the New Economy. DCMS, Crown Copyright.

Boje, D. M. (1991). 'The storytelling organization: A study of story performance in an office-supply firm.' *Administrative Science Quarterly* **36**: 106–126.

Boyle, J. (2008). 'The Public Domain: Enclosing the Commons of the Mind.' thepublicdomain.org.

Boyle, J. (2011). 'Presumed guilty.' *Financial Times*, 23rd February.

Burgelman, R. A. (1983). 'A model of the interaction of strategic behavior, corporate context, and the concept of strategy.' *Academy of Management Review* **8**: 61–70.

Burl, A. (2006). *Black Barty: The Real Pirate of the Caribbean*. Stroud, Sutton Publishing.

Burr, V. (1995). *Introduction to Social Constructionism*. London, Routledge.

Carr, N. (2010). *The Shallows: How the Internet Is Changing the Way We Think, Read and Remember*. New York, Atlantic Books.

Cheney, G., Lars Thoger Christensen, Charles Conrad, Daniel J. Lair (2004). 'Corporate rhetoric as organizational discourse.' *The Sage Handbook of Organizational Discourse*. D. Grant, C. Hardy, C. Oswick and L. L. Putnam. London, Sage, 79–103.

Coleman, M. (2005). *Playback: From the Victrola to MP3, 100 Years of Music, Machines, and Money*. Cambridge, MA, Da Capo.

Cyert, R. M. and J. G. March (1963). *A Behavioural Theory of the Firm*. Englewood Cliffs, NJ, Prentice Hall, republished by Blackwell, Oxford 1992.

David, M. and J. Kirkhope, Eds. (2005). *The Impossibility of Technical Security: Intellectual Property and the Paradox of Informational Capitalism*. Global Politics in the Information Age Manchester, Manchester, Manchester University Press.

Deazley, R., Martin Kretschmer, Lionel Bently (2010). *Privilege and Property: Essays on the History of Copyright*. Cambridge, Open Book Publishers.

De Bono, E. (1984). *Tactics: The Art of Science and Success.* Boston, Little, Brown.

Defoe, D. (1720). *The King of Pirates: Being an Account of the Famous Enterprises of Captain Avery, The Mock King of Madagascar.* eBook Project Gutenberg, Producer Jens Sadowski, 2011.

Defoe, D. (2005). 'Of captain mission.' R. Boys and R. Cohen. Project Gutenberg eBook.

De Wit, B. and R. Meyer (2005). *Strategy Synthesis: Resolving Strategy Paradoxes to Create Competitive Advantage.* London, Thomson.

Disco, C. and B. van der Meulen (1998). 'Introduction.' *Getting New Technologies Together.* C. Disco and B. van der Meulen. Berlin, De Gruyter, 1–13.

Dowd, T. (2006). 'From 78s to MP3s: The embedded impact of technology in the market for prerecorded music.' *The Business of Culture.* J. Lampel, J. Shamsie and T. K. Lant. Mahwah, New Jersey, Lawrence Erlbaum.

Earle, P. (2004). *The Pirate Wars.* London, Methuen.

Fairclough, N. (2001). 'The discourse of new labour: Critical discourse analysis.' *Discourse as Data.* M. Wetherell, S. Taylor and S. J. Yates. Milton Keynes, Open University/Sage.

Fisher, E. (2010). *Media and New Capitalism in the Digital Age: The Spirit of Networks.* New York, Palgrave Macmillan.

Fisher III, W. W. (2004). *Promises to Keep: Technology, Law, and the Future of Entertainment.* Stanford, CA, Stanford University Press.

Forster, E. M. (1927). *Aspects of the Novel.* London, Edward Arnold.

Foucault, M. (1969). *What Is an Author* (trans. Josue V. Harari). wiki.brown.edu/confluence/download/attachments/74858352/FoucaultWhatIsAnAuthor.pdf.

Foucault, M. (1977). 'Nietzsche, genealogy, history.' *Language, Counter-memory, Practice: Selected Essays and Interviews.* D. Bouchard. New York, Cornell University Press.

Foucault, M. (1980). *Power/Knowledge: Selected Interviews and Other Writings.* New York, Pantheon.

Gabriel, Y. (2004). 'Narratives, stories and texts.' *The Sage Handbook of Organizational Discourse.* D. Grant, C. Hardy, C. Oswick and L. L. Putnam. London, Sage.

Gantz, J. and J. Rochester (2005). *Pirates of the Digital Millennium: How the Intellectual Property Wars Damage Our Personal Freedoms, Our Jobs, and the World Economy.* New Jersey, FT Prentice Hall.

Gelatt, R. (1955). *The Fabulous Phonograph: The Story of the Gramophone from Tinfoil to High Fidelity.* London, Cassell.

Goodell, J. (2003). Steve Jobs: The Rolling Stone Interview. *Rolling Stone.*

Gore, A. (2013). *The Future.* New York, W H Allen.

Grant, D., Cynthia Hardy, Cliff Oswick, Linda L. Putnam (2004). *The Sage Handbook of Organizational Discourse.* London, Sage.

Greimas, A. (1987). *On Meaning: Selected Writings* (trans. Perron & Collins). Minneapolis, University of Minnesota Press.

Griffin, J. (2008). 'OneHouse LLC – Jim Griffin's Bio.' http://www.onehouse.com/bio.htm.

Hardin, G. (1968). 'The tragedy of the commons.' *Science* **162 (3859)**: 1243–1248.

Hargreaves, I. (2011). Digital Opportunity: A Review of Intellectual Property and Growth. Independent Report for the UK Government.

Hendry, J. (2000). 'Strategic decision making, discourse, and strategy as social practice.' *Journal of Management Studies* 37 (7): 955–977.

Hesmondhalgh, D. (2002). *The Cultural Industries*. London, Sage.

Hesmondhalgh, D. (2013). *The Cultural Industries* (3rd Edition). London, Sage.

Huygens, M., Frans A. J. Van Den Bosch, Henk W. Volberda, Charles Baden-Fuller (2001). 'Co-Evolution of firm capabilities and industry competition: Investigating the music industry, 1877–1997.' *Organization Studies* 22: 971.

Ingram, J. R. and S. Hinduja (2008). 'Neutralizing music piracy: An empirical examination.' *Deviant Behavior* 29 (4): 334–366.

Jefferson, T. (1813). 'Letter to Isaac McPherson, August 13th. Monticello.' founders.archives.gov.

Jobs, S. (2007). 'Thoughts on Music.' www.apple.com/hotnews/thoughts onmusic/.

Johns, A. (2009). *Piracy. The Intellectual Property Wars from Gutenberg to Gates*. Chicago, Univesrity of Chicago Press.

Johnson, C. C. (1998). *The General History of the Robberies and Murders of the Most Notorious Pyrates, and also Their Policies, Discipline and Government*. London, Conway.

Katz, M. (2004). *Capturing Sound: How Technology Has Changed Music*. Berkeley, CA, University of California Press.

Keightley, K. (2004). 'Long play: Adult-Oriented popular music and the temporal logics of the post-war sound recording industry in the USA.' *Media Culture Society* 26: 375.

Kelly, K. (2008). 'Better than free.' *The Technium*, 31st January 2008. http://www.kk.org/thetechnium/.

Knights, D. and G. Morgan (1991). 'Corporate strategy, organizations and subjectivity: A critique.' *Organisation Studies* 12 (2): 251–273.

Kontorovich, E. (2004). 'The piracy analogy.' *Harvard International Law Journal* 45 (1): 183–237.

Kuhn, M. (2002). *One Hundred Films and a Funeral*. London. Thorogood.

Kusek, D. and G. Leonhard (2005). *The Future of Music: Manifesto for the Digital Music Revolution*. Boston, MA, Berklee Press.

Lanier, J. (2013). *Who Owns the Future?* London, Allen Lane.

Lefsetz, B. (2008). 'The Lefsetz Letter (19th June 2008).' http://lefsetz.com/wordpress/archives.php.

Lessig, L. (2002). 'Keynote from OSCON.' Retrieved 21 June 2007, www.oreillynet.com/pub/a/policy/2002/08/15/lessig.html.

Lessig, L. (2004). *Free Culture: How Big Media Uses Technology and the Law to Lock Down Culture and Control Creativity*. New York, Penguin.

Lewis, S. (2012). 'The myth of the tragedy of the commons.' *The Wild Peak*. www.thewildpeak.wordpress.com.

Luckombe, P. (1771). *The History and Art of Printing*. London, Adlard & Browne for J. Johnson.

Maturana, H. R. and F. J. Varela (1980). *Autopoiesis and Cognition: The Realization of the Living*. Dordrecht, Holland, Reidel.

McLuhan, M. (2011). *The Gutenberg Galaxy*. Toronto, University of Toronto Press.

Miege, B. (1989). *The Capitalization of Cultural Production*. New York, International General.

Milton, J. (2006). *Areopagitica: A Speech for the Liberty of Unlicensed Printing to the Parliament of England, 1644.* Project Gutenberg eBook.

Mintzberg, H. (1994). *The Rise and Fall of Strategic Planning.* New York, Free Press.

Mintzberg, H., Bruce Ahlstrand, Joseph Lampel (1998). *Strategy Safari.* London, Pearson Education.

Morozov, E. (2011). *The Net Delusion: How Not to Liberate the World.* London, Penguin.

Munir, K. A. and N. Phillips (2005). 'The birth of the "Kodak Moment": Institutional entrepreneurship and the adoption of new technologies.' *Organization Studies* **26**: 1665–1687.

Murdock, G. and P. Golding (1999). 'Common markets: Corporate ambitions and communication trends in the UK and Europe.' *The Journal of Media Economics* **12** (2): 117–132.

Nicoli, E. (2007). 'EMI Press Release.' http://www.emigroup.com/Press/2007/press18.htm.

Nonaka, I. and H. Takeuchi (1995). *The Knowledge-Creating Company.* New York, Oxford University Press.

Ong, W. (1982). *Orality and Literacy: The Technologizing of the Word.* London, Methuen.

Pallante, M. (2013). The Next Great Copyright Act. *26th Horace S. Manges Lecture.* Columbia Law School.

Pallante, M. (2013) Q&A with Glenn Peoples. 18th December 2013. "Billboardbiz. www.billboard.com"

Parish, J. R. (1995). *Pirates and Seafaring Swashbucklers on the Hollwood Screen: Plots, Critiques, Casts and Credits for 137 Theatrical and Made-for-Television Releases.* Jefferson, NC, McFarland & Co.

Parker, M. (2009). 'Pirates, merchants and anarchists: Representations of international business.' *Management & Organizational History* **4** (2): 18.

Pascale, R. T., Mark Millemann, Linda Gioja (2000). *Surfing the Edge of Chaos: The Laws of Nature and the New Laws of Business.* New York, Three Rivers Press.

Pearson, K. A. and D. Large, Eds. (2005). *The Nietzsche Reader.* Oxford, Wiley-Blackwell.

Peifer, K.-N. (2010). 'The return of the commons.' *Privilege and Property.* R. Deazley, M. Kretschmer and L. Bently. Cambridge, Open Book Publishers.

Pettigrew, A. (1977). 'Strategy formulation as a political process.' *International Studies of Management & Organization* **7** (2): 78–87.

Pettigrew, A. M. (1992). 'The character and significance of strategy process research.' *Strategic Management Journal* **13**: 5–16.

Porter, M. E. (1985). *Competitive Advantage: Creating and Sustaining Superior Performance.* New York, Free Press.

Potter, D. (2012). 'Face to face with Dennis Potter (1986).' *Sabotage Times,* www.sabotagetimes.com/people/face-to-face-with-dennis-potter/

Potter, J. and M. Wetherell (1987). *Discourse and Social Psychology: Beyond Attitudes and Behaviour.* London, Sage.

Pyle, H. (2006). *Howard Pyle's Book of Pirates.* Project Gutenberg eBook, Compiled by Merle Johnson.

Reyman, J. (2010). *The Rhetoric of Intellectual Property.* New York, Routledge.

Rideau, F. (2008). Commentary on Diderot's *Letter on the Book Trade*. *Primary Sources on Copyright (1450–1900)*. L. Bently and M. Kretschmer. www.copyrighthistory.org.

Ringer, B. (1974). *The Demonology of Copyright*. New York, R.R. Bowker Company.

Rosati, E. (2013). 'The Orphan Works Provisions of the ERR Act: Are They Compatible with UK and EU Laws?' *papers.ssrn.com*.

Rose, M. (2010). 'The public sphere and the emergence of copyright.' *Privilege and Property*. M. Kretschmer, R. Deazley and L. Bently. Cambridge, Open Book Publishers.

Rumelt, R. (2011). *Good Strategy/Bad Strategy*. London, Profile Books.

Schendel, D. E. and C. H. Hofer, Eds. (1979). *Strategic Management: A New View of Business Policy and Planning*. Boston, Little, Brown.

Senge, P., Art Kleiner, Charlotte Roberts, Richard Ross, Bryan Smith (1994). *The Fifth Discipline Fieldbook*. London, Nicholas Brealey.

Simon, H. A. (1957). *Models of Man*. New York, Wiley.

Sousa, J. P. (1906). 'The menace of mechanical music.' *Appleton's Magazine* **8**: 278–284.

Spitz, D. and S. Hunter (2005). 'Contested codes: The social construction of napster.' *The Information Society* **21**: 169–180.

Sterne, J. (2012). *MP3 the Meaning of a Format*. Durham, NC, Duke University Press.

Surowiecki, J. (2005). *The Wisdom of Crowds: Why the Many Are Smarter than the Few*. London, Abacus.

Sykes, G. M. and D. Matza (1957). 'Techniques of naturalization: A theory of delinquency.' *American Sociological Review* **22** (6): 664–670.

Taylor, J. R. and F. Cooren (2006). 'Making worldview sense: And paying homage, retrospectively, to Algirdas Greimas.' *Communication as Organizing: Empirical and Theoretical Explorations in the Dynamic of Text and Conversation*. F. Cooren, J. R. Taylor and E. J. Van Every. Mahwah, New Jersey, Lawrence Erlbaum.

Turow, S., Paul Aiken, James Shapiro (2011). 'Would the bard have survived the web?' *New York Times*.

Twenge, J. (2006). *Generation Me*. New York, Free Press.

Wallas, G. (1926). *The Art of Thought*. New York, Harcourt Brace.

Weick, K. E. (2001). *Making Sense of the Organization*. Malden, MA, Blackwell.

Wilde, O. (1891). *The Picture of Dorian Gray*. London: Ward, Lock & Co.

Williams, O. (1923). 'Times and seasons.' *Gramophone* **1**: 38–39.

Williamson, J. and M. Cloonan (2007). 'Rethinking the music industry.' *Popular Music* **26** (2): 305–322.

Willig, C. (2001). *Introducing Qualitative Research in Psychology: Adventures in Theory and Method*. Buckingham, Open University Press.

Index

Abbey Road studios, 32
Access to Knowledge (A2K), 208
  see also public domain
Adorno, Theodor, 234
advertising-supported model, 84–6
AFM strike, 33–4
  see also Petrillo, James Caesar
Alexander the Great, 199
Amazon.com, 24, 53, 58, 226
Anderson, Chris, 235
  see also *Wired* magazine
AOL Time Warner, 27, 39
Apple Inc, 24, 145, 167, 173–4, 183
  acquisition of Beats Electronics,
    52–3
  compatibility of products with
    MP3, 46
  digital rights management (DRM),
    105–7
  as entrepreneur-inventor, 158–60
  fear of dominance by, 58, 225–6
  launch of iPod and iTunes, 49–51
  music pricing, 100–1; *see also*
    unbundling; cherry-picking
  music strategy as *honey-trap*, 73, 141
  scepticism towards subscription
    services, 52, 81–2
  see also Jobs, Steve
Areopagitica (John Milton), 192,
  218–19
artificial scarcity, 11, 203, 214, 241
  see also copyright
ArtistShare, 146
  see also crowd funding; societies
AT&T, 42, 45, 103
  see also Bell Laboratories
Atkyns, Richard, 193–4
author rights (origins of), 193, 214–15,
  219, 221
  see also moral rights
Authors Guild v. Google, 229–231

Baby Boomers, 50, 181
  see also Generation X; Generation Y
Barlow, John Perry, 209–10
  see also Electronic Frontier
    Foundation
Barraud, Francis, 30
Barthes, Roland, 18, 222
battle of the booksellers, 216, 219
Beats Electronics, 52–3, 81–2
Beecham, Thomas, 32, 41
Bell, Alexander Graham, 29
Bell Laboratories, 31, 42
  see also compression (digital audio)
Berliner, Emile, 29–30, 38
Berne Convention (1886),
  216, 221
Bertelsmann Music Group, *see* BMG
Black Barty (Captain Bartholomew
  Roberts), 200
BMG (Bertelsmann Music Group), 27,
  38, 77, 136, 176–7
book publishing, 3, 57, 185, 192–4,
  206, 208, 215–31, 235, 243
  see also author rights; Authors Guild
    v. Google; battle of the
    booksellers; Caxton; Gutenberg;
    intellectual property; linear
    literary mind; public domain;
    orality; Stationers' Company
Boomers, *see* Baby Boomers
Boonstra, Cor, 47
  see also Philips; Timmer, Jan
bootlegging, 187
  see also piracy
Boyle, James, 18, 203, 205–6,
  211, 231
BPI, the, 172, 187
Brandenburg, Karlheinz, 160
  see also Fraunhofer Institute; Motion
    Pictures Expert Group
Bronfman, Edgar Jr, 26, 38–9
Brunel, Isambard Kingdom, 221